AF576011

Volume II

Volume II: Low Enthalpy Geothermal Energy

Special Issue Editors

Rajandrea Sethi

Alessandro Casasso

MDPI • Basel • Beijing • Wuhan • Barcelona • Belgrade • Manchester • Tokyo • Cluj • Tianjin

Special Issue Editors
Rajandrea Sethi
Politecnico di Torino
Italy

Alessandro Casasso
Politecnico di Torino
Italy

Editorial Office
MDPI
St. Alban-Anlage 66
4052 Basel, Switzerland

This is a reprint of articles from the Special Issue published online in the open access journal *Energies* (ISSN 1996-1073) (available at: https://www.mdpi.com/journal/energies/special_issues/vol_II_geothermal_energy).

For citation purposes, cite each article independently as indicated on the article page online and as indicated below:

LastName, A.A.; LastName, B.B.; LastName, C.C. Article Title. *Journal Name* **Year**, *Article Number*, Page Range.

ISBN 978-3-03936-284-4 (Pbk)
ISBN 978-3-03936-285-1 (PDF)

Cover image courtesy of Dr. Marco Orsi
geologist, http://geolorsi.it

Contents

About the Special Issue Editors

Rajandrea Sethi is a Full Professor and Head of the Groundwater Engineering Research Group at Politecnico di Torino. From 2015 to 2019 he served as Head of the Department of Environment, Land and Infrastructure Engineering. He has supervised more than 100 MSc candidates and 13 PhD students. He serves as an Associate Editor of Water Resources Research and has authored more than 70 journal articles focused on groundwater remediation, flow and contaminant transport, and geothermal energy. He is the author of the textbooks: Groundwater Engineering, published by Springer, and Nanomaterials for the Detection and Removal of Wastewater Pollutants, published by Elsevier.

Alessandro Casasso is an Assistant Professor at Politecnico di Torino, Department of Environment, Land and Infrastructure Engineering. His research focuses on shallow geothermal energy, covering aspects such as subsurface heat transport, environmental benefits and impacts, energy demand and efficiency, policy and regulation. He also addresses other aspects of groundwater protection, collaborating in expert consultancies on landfills, contaminated sites and large construction works. He has supervised more than 20 MSc candidates and is the author of 20 Scopus-indexed articles.

Preface to "Volume II: Low Enthalpy Geothermal Energy"

"Low enthalpy" geothermal energy is defined as the thermal energy available in the ground at temperatures below 90 °C and, generally, at depths below 400 m. At such a low temperature, this resource cannot be conveniently exploited to produce electricity, but it can be devoted to a wide range of thermal uses. It is most frequently used for heating and cooling of buildings, but also for a few industrial processes, including road de-icing, and bathing.

Research on how to exploit low enthalpy geothermal energy is published at an increasing rate. The main subjects covered by these works are technological development of single components (heat pumps, heat exchangers, wells etc.) and their integration, mathematical and numerical methods for heat transport modeling and the design of geothermal systems, assessment of shallow geothermal potential, technological transfer initiatives, study of economic aspects, and assessment of environmental impacts (both positive and negative).

We guest-edited this Special Issue in the year 2019, gathering seven articles covering some of these topics.

Perego et al. present a techno-economic assessment of the shallow geothermal energy potential in a Swiss canton (Cantone Ticino). This article also provides key insights on the management of the spatial density of shallow geothermal systems.

Sáez Blázquez et al. present a comparative analysis of different methodologies to estimate ground thermal conductivity: needle probe measurements, geophysical surveys, and thermal response tests, with this last method considered the benchmark for comparison.

Łukawska et al. present the methodology for laboratory thermal conductivity measurements of recompacted samples of cohesive and non-cohesive soils.

Kurevija et al. present the results of tests conducted on special borehole heat exchanger pipes with ribbed inner walls, which increase the turbulence of the heat carrier fluid and, thus, reduce borehole thermal resistance.

Jaziri et al. present the monitoring of a shallow geothermal system composed of 31 borehole heat exchangers operating with high velocity groundwater flow due to the dewatering of a nearby quarry.

Licharz et al. present the monitoring of two groundwater heat pump systems installed for the heating and cooling of pig houses.

Çuhac et al. present work on the possibility of harvesting heat from asphalt with specifically designed heat exchangers.

We would like to thank all the authors for their interesting, high-quality articles and the reviewers for their anonymous and invaluable work.

Rajandrea Sethi, Alessandro Casasso
Special Issue Editors

Article

Techno-Economic Mapping for the Improvement of Shallow Geothermal Management in Southern Switzerland

Rodolfo Perego [1,*], Sebastian Pera [1] and Antonio Galgaro [2]

1 Institute of Earth Sciences, University of Applied Sciences and Arts of Southern Switzerland SUPSI, Campus Trevano, CH-6952 Canobbio, Switzerland; sebastian.pera@supsi.ch
2 Department of Geosciences, University of Padova, Via Gradenigo, 6, 35131 Padova, Italy; antonio.galgaro@unipd.it
* Correspondence: rodolfo.perego@supsi.ch; Tel.: +41-58-666-62-18

Received: 13 December 2018; Accepted: 11 January 2019; Published: 16 January 2019

Abstract: Cantone Ticino, a mountainous region located in the southern part of Switzerland, is greatly affected by the continuous growth of subsurface exploitation through the use of both closed-loop and open-loop geothermal systems. In this study, techno-economic maps for shallow geothermal potential of Cantone Ticino are produced, considering closed-loop systems. The work starts with the identification of the main parameters affecting the techno-economic potential such as GST and thermal conductivity. Maps for different indicators of techno-economic feasibility are created and compared against real data/measurements. An empirical method is tailored to derive a map of the techno-economic geothermal potential, expressed as meters required to provide 1 kW of installed power. The produced map shows an overall discrepancy from real installed length data of approximately ±23%. Moreover, compared with current regulation, the produced maps show an unoptimized management of the shallow geothermal resource, since high potential zones are commonly located where the installation of BHE is not permitted and often closed-loop systems are installed where the estimated potential is lower, mainly in alluvial fans. In light of these considerations, the authorization process in Cantone Ticino for BHE should be revised taking into account the real techno-economic potential.

Keywords: Ground Source Heat Pumps (GSHP); mapping; potential; Switzerland; payback period; shallow geothermal

---

1. Introduction

Low enthalpy geothermal energy is a renewable energy that is becoming widely exploited within Europe, especially through the use of closed-loop systems. These systems exchange heat with the subsurface by means of a thermo-vector fluid, which circulates in a plastic pipe installed in the subsurface. It is a safe technology which proved to be very efficient and economically advantageous during the years given a proper design and installation. This technology is relatively mature but it is not cheap, since the installation costs are the main constraints that reduce its deployment throughout the entire Europe and economic benefits are noticed after years of operation.

Switzerland is one of the most advanced countries from the low-enthalpy geothermal energy standpoint. The state-of-the-art drilling techniques, the favorable geological and thermal conditions, the consistent know-how owned by the sector experts make this technology one of the most popular between all the available renewable energies, with the highest spatial density of probes in the world [1].

According to the Swiss Federal Office of Energy [2] a large-scale use of geothermal heat pumps could allow obtaining a relevant reduction of CO_2 emissions and fossil fuels consumption. Geothermal heat pumps are usually implemented in new buildings, especially public ones that adopt the Swiss certification MINERGIE® [3], but this technology could also be used to satisfy energy demand in historical or cultural buildings.

Low enthalpy geothermal technology has already been well-deployed in Cantone Ticino, which is a region located in southern Switzerland, delimited to the North by Gotthard Massif and surrounded by Italy to its West (Cantone Vallese, Lombardy and Piedmont), South (Lombardy) and East (Lombardy and Cantone Grigioni) (Figure 1).

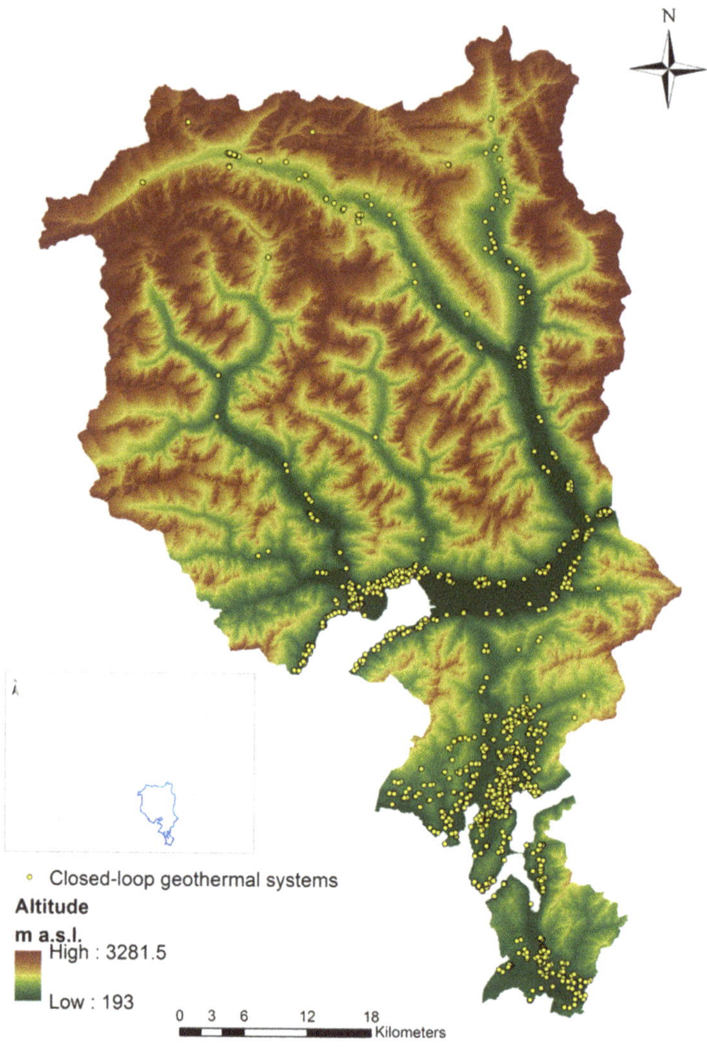

Figure 1. Location of the study area, with highlighted the distribution of closed-loop borehole heat exchanger (BHE) heat exchangers.

The last two decades (approximately from 1997 to 2017) have led to the current massive subsurface thermal exploitation by means of both closed-loop and open-loop geothermal systems. As a consequence, the region currently hosts a large number of geothermal installations, with an overall density of approximately 1.5 probes/km^2. Considering only main aquifers, where the majority of population lives, this density increases to 16 probes/km^2. Considering main cities, the value rises to 44 probes/km^2. This large amount of installations subsequently arise issues regarding short mutual distances or adjoining probe fields that will influence ground temperatures and system performances in the long term [4]. Shallow geothermal energy development in Cantone Ticino will be increasingly important, since at least 20% of energy requirements for new buildings will have to be provided from renewable energies as stated in [5]. Currently Cantone Ticino uses 28.9% of produced energy for building conditioning [6]. According to 2016 data, the annual energy demand for residential buildings is 3.398 GWh, equivalent to 35.6% of Cantonal consumption. Approximately 80% of this energy is used for building conditioning (2760 GWh in 2016) and 20% for electrical devices and lighting. The main energy source used for heating is heating oil (55.5%). The other sources are: natural gas (18%), electricity (11.9%), wood (7.6%) and, combined, solar thermal, heat pumps and heat from waste (7%). As a rough estimation, CO_2 emissions produced by residential buildings in Cantone Ticino are quantified in 562,000 tCO_2/year, of which 449,600 tCO_2/year are emitted from heating.

1.1. Closed-Loop Systems in Cantone Ticino Diffusion And Licensing Process

In 1997, less than 10 probes were installed and registered, according with [7], while by 2005 this number had grown to more than 2000. During the last 12 years, from 2004 to 2016, requests for new geothermal system permits (both closed and open, small to large systems) went from 30 per year to more than 100 per year, with peaks of 140 per year in 2011 and 2012. Currently new requests for thermal use of the underground are a regular occurrence: the information is therefore stored in a specific database. The SUPSI- Institute of Earth Sciences (IST)'s geodatabase GESPOS (Surveys, Wells and Springs Management) [8] hosts more than 4300 georeferenced and indexed closed-loop geothermal probes distributed in more than 1100 installations. For each installation a series of administrative and technical information is collected. Data analysis allows estimating an overall closed-loop installed power of approximately 30 MW and an aggregate installed borehole length of approximately 550 km, proving a strong deployment of this technology.

The majority of GSHP systems are located within Quaternary deposits (43%) filling glacial valleys, which constitute the main background for human activities and energy exploitation. A consistent amount of probes is also installed in gneiss (33%), granite (10%), limestone and green schist (6%). Figure 2a shows the frequency of each system type classified for its size: the most frequent types of installations are constituted of one and two probes (respectively 27.6% and 30% of the total). Systems constituted by more than 10 probes represent only 7.5% of the total, but they account for 35% of the overall number of probes.

Figure 2b presents the average total power per number of probes in the system. The weighted mean heating capacity observed in Cantone Ticino is approximately 25 kW. A power regression between installation power below 100 kW and probes number was found (R^2 = 0.956): according to this regression, for a mean heating capacity of 25 kW a representative system was constituted of four probes. This scenario was considered as referential for further simulations presented in this paper.

The authorization process of closed-loop systems is based on restrictions arising from the enforcement of the Swiss water protection act and ordinance [9] and contained in [10]: an excerpt of this regulation is shown in Table 1. It states that new closed-loop systems cannot be installed within S1 and S2 wellhead protection zones and as a rule they are not allowed even in S3, except for specific situations. They are always allowed in üB ("übriger Bereich") remaining territory where there is no occurrence of exploitable groundwater. Within the Au (exploitable groundwater) sector, generally geothermal probes are not allowed except for "sacrifice areas" where the presence of conflicts precludes groundwater exploitation for drinking purposes [11]. In these areas the great presence of conflicts (sewers, industries etc.) precludes a creation of groundwater protection areas in compliance with the regulation: any thermal use of the underground is therefore admitted.

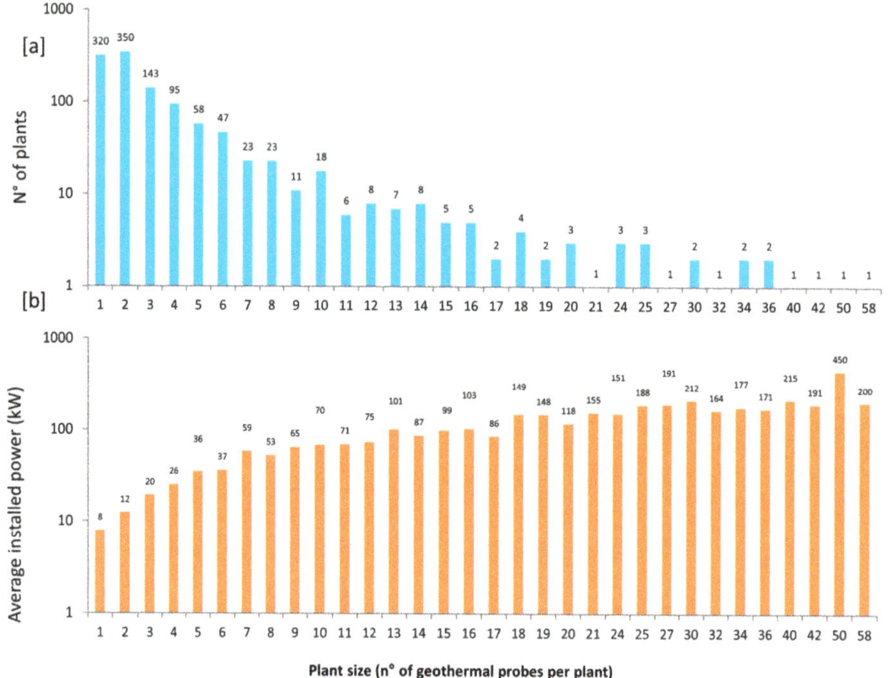

Figure 2. Basic statistics for Cantone Ticino geothermal systems: (**a**) amount of systems per size class; (**b**) average installed power per size class.

Table 1. Reference table for the exploitation of heat from ground and subsurface in Switzerland.

Type of System	üB	Au	Zu	Area	S3	S2	S1
Geothermal probes (vertical systems)	+	b		−	−	−	−
Underground circuits (horizontal systems)	+	+[4]		−[2/4]	−[b/5/4/7]	−	−
Geothermal piles and other thermo-active elements	+	b		−[2/4]	−[b]	−	−
Wells for groundwater withdrawal, for heating and cooling	+	b		−	−	−	−
Coaxial wells	−[6]	−[6]		−	−	−	−

(+) No restrictions; (b) admitted or not after a case by case analysis by the Authority, requires authorization; (−) forbidden; (−b) forbidden but could be waived if it is a particular case; (2) can be authorized if a detailed hydrogeological report identifies the future limits of the S3 protection zones; (4) the installations must be realized at least 2 m above the maximum piezometric head level; (5) no direct-expansion heat pump; (7) if circuits are located in the soil (horizons A or B) and not in the sub-soil, an authorization can be granted; (6) authorized only if specific geophysics and hydrogeological studies state that there is no threat for groundwater.

A case-by-case analysis of environmental impact assessment for every GSHP system can be a difficult, time-consuming and expensive task, so a detailed and verified map of the techno-economic potential can be useful. Cantone Ticino adopts a "rule of thumb" to prevent potential interferences among neighboring installations: new systems must be installed at a distance of at least 5 m from the edges of the parcel they are installed in. In this way adjacent parcels should have two neighboring geothermal systems which are at least at 10 m distance, ideally preventing from potential thermal interferences. The estimation of the suitability for closed-loop GSHPs and its mapping is important from an economic and decisional standpoint, since it could give precious advices on how to regulate its deployment and could allow a better management of thermal resource. The aim of this paper is to give some indications to improve the current authorization procedure of GSHP systems, by adopting more dynamic and physically robust methods which also account for the extractable potential. This approach would better balance environmental, technological and legal aspects on a spatial basis. This procedure is achievable by using Geographic Information Systems (GIS), which are powerful tools that can manage numerous and large spatial data, performing computationally demanding calculations. The aim was to obtain quality products with high spatial resolution even for large territories.

1.2. Low Enthalpy Geothermal Potential Mapping in Literature

The scheme proposed by the United States National Renewable Energy Laboratory (NREL) [12], summarizes a clear concept of "potential" referred to a renewable energy. The technical potential of a renewable energy is stated as the "achievable energy generation of a particular technology given different constraints". At the base of the pyramid there is the physical theoretical energy extractable or generated by the renewable source, which is always limited by the technological means and by the expertise. In particular, for shallow geothermal systems, the extractable energy is limited by the COP/EER of the heat pump, by probe material and grouting. Technical potential is limited by economic potential, since a high-performance technology could be economically disadvantageous due to high fuel consumption, high material production costs or high labor costs. The top of the pyramid is represented by market constraints, which commonly represent policy implementation, regulatory restrictions, competition with other technologies, investor response and the attractiveness of the resource perceived by stakeholders. In particular, this last aspect has become more and more important in the last years to promote renewable energy solutions.

In the past decades few research papers focused on the estimation of shallow geothermal potential, especially through mapping procedure. Only recently this topic has gained visibility, clearly proving that the spatial management of shallow geothermal systems (both open and closed) is becoming more relevant at local or regional scale.

A shortlist of papers focusing on this topic could begin with the work from [13], which gives suggestions on the parameters to consider for the assessment of the potential, such as the percentage of heat demand that could be satisfied by geothermal systems, maximum allowable drilling depth and potential reduction of CO_2 emissions.

The work from Garcia-Gil et al. [14] uses analytical formulas to calculate the Low Temperature Geothermal Potential (LTGP) for both open and closed systems. The mapping work from Bertermann et al. [15] estimates the Very Shallow Geothermal Potential (vSGP) for horizontal systems. A great amount of data coming from different scientific areas had to be homogenized and implemented into an online GIS. The work from Gemelli et al. [16] proposes a methodology to map economic factors affecting GSHP systems such as energy cost, payback time, €/kg, CO_2.

The work from Casasso et al. [17] estimates both open-loop and closed-loop potential for the Province of Cuneo (Italy), calculating the amount of extractable energy from the underground, by means of a specifically tailored analytical equation. The work from Arola et al. [18] estimates the open-loop potential of Finnish aquifers based on both physical properties of aquifers and the efficiency of heat pumps. The paper from Viesi et al. [19] proposes a comprehensive work that is aimed at deriving the geo-exchange potential in the Adige Valley (Italy), by using a large amount of collected data, including hydrogeological, climate and lithostratigraphic information.

Detailed work at small-scale by Galgaro et al. [20] estimates the geo-exchange potential through a comprehensive characterization of the subsurface and of the energy request from a reference residential unit.

In this paper the required BHE length that could satisfy an assigned energy requirement was considered to be the minimum index that could better describe both natural and technological constraints. From this indicator it was also possible to estimate the capital cost of the GSHP installation, which in turn allows producing economical and market index maps. The modeling part of this paper is inspired by the work from Galgaro et al. [20], especially the use of EED software [21] and the empirical relations between 3D parameters.

The work procedure followed both an empirical and a regulatory/standard approach: in particular, some formulas used to evaluate temperature values (e.g., ground temperature) and energy demand parameters (e.g., the annual heating demand index) are taken from Swiss regulations/standards, such as SIA 384/6 [22] and MINERGIE®, in order to have reference values that could be consistent for Switzerland. Also most of thermal conductivity values are taken from SIA 348/6 regulation, since few in-situ thermal measurements were available. The presented work forms part of the HORIZON2020 "Cheap-GSHPs" project, which aims at reducing the installation costs of closed-loop shallow geothermal systems. The work performed by SUPSI (partner of the consortium) on Cantone Ticino was useful to test at European level a mapping procedure to estimate the techno-economic potential (for the description of the received funding see "Funding" section).

2. Natural Resource

The mapping procedure started with the identification of the main resource parameters affecting vertical closed-loop GSHPs operation, efficiency and dimensioning. For the considered study area, the investigated natural parameters were mean annual air temperature (MAAT) of the location, that affects ground surface temperature (GST), and thermal conductivity of the subsurface. Thermal conductivity was calculated separately for both outcrops and unconsolidated material, since they require completely different methodologies.

2.1. Mean Annual Air Temperature (MAAT) and Ground Surface Temperature (GST) Mapping

The initial part of the work involved the creation of a Mean Annual Air Temperature (MAAT) map for Cantone Ticino that could allow estimating Ground Surface Temperature (GST), a required input for EED software. In fact GST is used by the software to calculate ground temperature at half of the borehole length, using both thermal conductivity and heat flux information. 30 years climatic normal of GST direct measurements were not available for Cantone Ticino nor for Switzerland. A robust and homogeneous air temperature database was retrieved from the MeteoSwiss [23]. MeteoSwiss stations located within Cantone Ticino (with the exception of S. Bernardino and Grono ones) with annual average values measured at a distance of 2 m above ground and homogenized over the 1981–2010 period were considered for the development of the MAAT map. Table 2 shows the location of the used monitoring stations and the corresponding MAAT values.

Table 2. Information regarding Cantone Ticino's MeteoSwiss stations used for Mean Annual Air Temperature (MAAT) map creation.

Station Name	Spatial Information			Normal Period: 1981–2010 (Datum: WGS 84)
	Latitude (Decimal Degrees)	Longitude (Decimal Degrees)	Altitude (m a.s.l.)	Measured MAAT (°C)
Lugano	46.0042	8.9603	273	12.4
Stabio	45.8434	8.9323	353	11.1
Cimetta	46.2004	8.7916	1661	5.2
Locarno Monti	46.1724	8.7875	367	12.4
Magadino/Cadenazzo	46.1600	8.9336	203	11.4
Grono	46.2550	9.1637	324	12.4
Acquarossa/Comprovasco	46.4595	8.9354	575	9.9
Piotta	46.5148	8.6880	990	7.7
S.Bernardino	46.4635	9.1846	1639	3.9

The mapping procedure was performed at a small region-scale, so only the effects of altitude were taken into account without considering latitude. A linear regression between altitude data and temperature data was performed, finding a regression formula with a Pearson R^2 factor of 0.95

that led to an error of approximately 7.4% (0.65 °C) between measured and estimated temperature. The discovered linear equation was:

$$\text{M.A.A.T. } (^\circ\text{C}) = \frac{\text{Altitude} - 2385.7}{-174.61} \tag{1}$$

The formula was applied to a 25 m Digital Elevation Model (DEM), obtaining the MAAT map for Cantone Ticino (Supplementary Material Figure S1).

As expected, valleys show higher mean annual air temperatures and this reflects on annual ground temperatures. In literature, as a good approximation, MAAT is commonly considered as equal to the undisturbed ground temperature [19,20]. Ground temperatures maps are therefore usually derived by creating MAAT maps. To understand if this assumption could be valid for an Alpine region like the studied one, MAAT map was compared with undisturbed ground temperatures obtained from Thermal Response Tests (TRTs) executed in the region between 2010 and 2015 at different altitudes. For this specific case the absolute average shift between MAAT and undisturbed ground temperature was quantified in 2.3 °C as seen in Table 3.

Table 3. Comparison between undisturbed ground temperatures from thermal response tests (TRTs) and MAAT.

Site Name	Lat. (DD, WGS84)	Long. [DD, WGS 84]	Altitude (m a.s.l.)	Undisturbed Ground Temperature from TRT (°C)	Estimated MAAT from Map (°C)	Absolute T Discrepancy (°C)
Barbengo	45.9596	8.9197	283	14.4	12	2.4
Collina d'oro	45.9631	8.9083	526	11.8	10.7	1.1
Lugano-Besso	46.0094	8.9380	378	14.8	11.5	3.3
Massagno	46.0115	8.9423	367	14.8	11.6	3.2
Mendrisio	45.8642	8.9824	356	12.7	11.6	1.1
Olivone (estimated coordinates)	46.5181	8.8842	1433	8.3	5.5	2.8
DATA SOURCE: various TRT executed in-situ between 2010 and 2015					MAE: 2.3 °C	RMSE: 2.5 °C

This proved that the produced MAAT map was not suitable to be used as a ground temperature map. The magnitude of the measured gap between MAAT and GST depends on snow cover and on the latent heat of fused ground moisture [24]. GST greatly affects the heat exchange between the probe and the surrounding ground, so a careful evaluation is required. A correction factor taken from Swiss regulation was therefore applied on the created MAAT map in order to derive a GST one. According to SIA 384/6, which is the Swiss regulation for closed-loop systems, the investigated region was then subdivided into two portions, depending on elevation classes. For a pixel elevation <1000 m, the equation used for GST estimate was:

$$\text{GST} = \text{MAAT} + 1.55\ ^\circ\text{C} \tag{2}$$

For an elevation >1000 m, the equation was:

$$\text{GST} = \text{MAAT} + 1.55\ ^\circ\text{C} + \frac{(\text{Altitude} - 1000)}{800} \times 2.45 \tag{3}$$

Results from the GST mapping procedure are shown in Figure 3. Five soil monitoring stations taken from [25] were used in order to compare the proposed reconstruction of GST with real measured data. Measured GST data consist of short time series of soil temperatures, aggregated monthly, at 0.1 m depth. Data comparison between mapped and measured GST is reported in Table 4.

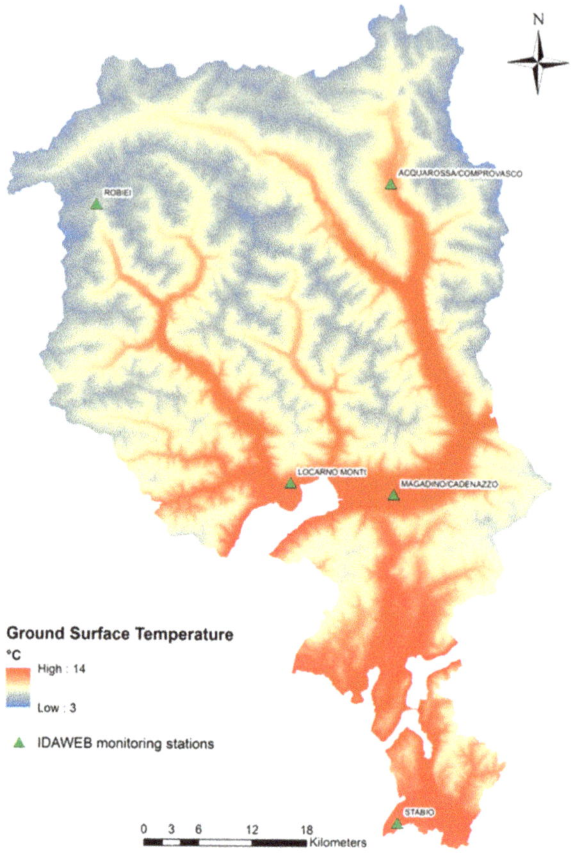

Figure 3. Map of reconstructed ground surface temperature (GST) with highlighted IDAWEB stations used for comparison between measured and estimated GST.

Table 4. Comparison between measured and estimated GST.

Site Name	Lat. (DD, WGS84)	Long. (DD, WGS84)	Altitude (m a.s.l.)	Consistency	Measured GST (°C)	Mapped GST (°C)	Absolute Discrepancy (°C)
Acquarossa/ Comprovasco	46.4594	8.9356	575	Average value for 3 years	10.6	11.9	1.3
Locarno monti	46.1725	8.7874	366	Average values for 9 years	13.6	13.1	0.5
Magadino/ Cadenazzo	46.1600	8.9336	203	Average value for 12 years	12.7	14.0	1.3
Robiei	46.4430	8.5133	1896	Average value for 6 years	5.1	7.1	2.0
Stabio	45.8433	8.9323	353	Average value for 4 years	11.8	13.2	1.4
		Data source: IDAWEB [25]				MAE: 1.3 °C	RMSE: 1.4 °C

The R^2 coefficient for the linear regression is 0.94, the overall discrepancy between measured and mapped GST can be defined as ±1.3 °C, with an RMSE of 1.4 °C. A second verification process was then performed: resulting GST map was compared with a robust approach proposed by Signorelli and Kohl (2004) for GST mapping calibrated for the entire Switzerland and valid for elevations ≤1500 m a.s.l. [24]. In the cited study the Swiss GST map was obtained by analyzing ground temperature data from 1984 to 2004 and determining GST dependence with altitude and with surface air temperature. A GST map was then built and verified against ground temperature values obtained by extrapolation from a large number of boreholes. Signorelli and Kohl's polynomial regression for elevations ≤1500 m a.s.l. was applied to the DTM:

$$GST = 15.23 - 1.08 \times 10^{-2} \times (\text{Elevation}) + 5.61 \times 10^{-6} \times (\text{Elevation})^2 - 1.5 \times 10^{-9} \times (\text{Elevation})^3 \quad (4)$$

Stations above 1500 m a.s.l. were not considered for this approach, since over this altitude the polynomial regression is not reliable according to what they state in [24]. This comparison showed that GST reconstruction of the present paper was not dissimilar from the polynomial regression of Signorelli and Kohl. Giving this double comparison process, the produced GST map was considered to be suitable for further elaborations. More measured data could surely improve the reliability of this GST reconstruction.

2.2. Subsurface Thermal Characterization

The subsequent step was devoted to the thermal characterization of the subsurface, which is of fundamental importance when estimating a techno-economic shallow geothermal potential. To derive a thermal conductivity map the territory was split into two separate portions, depending on material type: outcrops and unconsolidated materials within monitored aquifers. The thermal characterization of these two subsets required different mapping approaches.

2.2.1. Outcrops

Thermal conductivity map of outcrops was built by using a 1:500,000 scale geological map provided by [26]. Recommended values of thermal conductivity were assigned to each specific lithology according to SIA 384/6; in presence of multiple lithologies within the same polygon, an average value of recommended thermal conductivity was assigned. Dolomite and Vulcanites outcropping lithologies in southern Ticino were assigned thermal conductivity values according to [27], where λ measurements were performed in laboratory on field samples. The assigned thermal properties are reported in Table 5. Thermal conductivity map of outcrops was compared with five TRTs performed in predominant rock (Supplementary Material Table S1). Results show that the mapped estimated thermal conductivity is lower in almost all locations (average error = +1 W/mK). A thermal conductivity map at this scale cannot consider local anomalies, heterogeneities or local groundwater flow which affect (frequently improve) thermal properties of the subsurface. This could imply that the mapped estimated conductivities could be systematically underestimated and the required BHE slightly overestimated.

Table 5. Thermal characterization of outcrops and unmonitored Quaternary portion for in Cantone Ticino.

Lithology	λ (W/mK)	Data Source
Greenschists, amphibolites, metagabbro, meta-ultrabasite	2	Averaged from SIA 384/6
Pelitic and psammitic gneiss, fillads, conglomerates, sandstone	2.6	Averaged from SIA 384/6
Dolomite and dolomitic marble	3.17	Lab. measurements Soma, 2015
Granite, granodiorite	2.8	SIA 384/6
Acid and basic vulcanites	2.36	Lab. measurements Soma, 2015
Granite gneiss	2.7	SIA 384/6
Generic Quaternary deposits	2	Representative value SIA 384/6

2.2.2. Unconsolidated Materials

Only the main monitored aquifers were taken into account for the characterization of the unconsolidated portion: Magadino plain, Laveggio, Vedeggio, Valle Maggia and Chiasso. The used approach consisted in mapping the hydraulic conductivity and the shape of the groundwater table for each aquifer. The first reconstruction was useful to assign appropriate thermal properties to each zone, while the second reconstruction was useful to assess the depth of the groundwater.

The thickness of the vadose zone influences both the thermal conductivity of the subsurface and therefore the heat transfer rate as experimentally demonstrated in [28]. For monitored aquifers the shape of the water table between 2015 and 2017 was reconstructed by using mean annual hydraulic head data retrieved from a groundwater monitoring network composed of 110 monitoring points (10 automatic probes and approximately 100 piezometers/water wells). Groundwater data was used to estimate the thicknesses of both vadose zone and saturated portion considering a total of 100 m depth, which is commonly the typical depth of Borehole Heat Exchangers (BHEs). Using a wet/dry approach allows to more accurately describe the influence of vadose zone on λ values. This correction factor does not greatly affect zones with shallow groundwater but it is fundamental for deeper groundwater zones, such as alluvial fans.

The main assumption of the hydraulic conductivity map was that values obtained from pumping tests (in form of log10 k and the back-transformed k, where k is the hydraulic conductivity) could be considered as representative of all the subsurface column down to 100 m depth. This assumption is congruent with a mapping procedure for a 2800 km^2 region with output maps scale of approximately 1:400,000–1:500,000. Bedrock occurrences within 100 m depth were not considered, since not enough bedrock depth information (only few logs for all aquifers) was available. 556 hydraulic conductivity values measured from pumping tests were therefore analyzed and $\log_{10}$k was computed. If some wells showed more than one value of measured k (15 of 556) at different depths, logk averaging was performed. The computed hydraulic conductivity values were interpolated to create a continuous reconstruction. Ordinary kriging was chosen to interpolate between wells, since no trend within data was observed. The parameters used for both the realization of the experimental variogram and for the interpolation are reported in Table 6. The interpolated data were back-transformed using the k = $10^{\log k}$ relation. In order to assess the goodness and the reliability of the proposed reconstruction, two verification steps were performed. Firstly a leave-one-out cross-validation was performed using the same dataset used for the interpolation. Then a second dataset of measured logk was used to compare the results: this second dataset is constituted of 409 surveys distributed fairly homogeneously within the whole investigated area. Results of the comparison between the interpolated hydraulic conductivity and the second dataset are reported in Table 6.

Table 6. Ordinary kriging parameters (left) and cross-validation/comparison with real data (right) for hydraulic conductivity spatial reconstruction.

Variogram		Double Comparison Statistics		
Parameter	Value		Cross Validation	Surveys Data
Nugget	0.247	N° of data	555	409
Range (m)	1514	Max Abs error ($\log_k$)	3.51	2.26
Sill (Partial)	4.08 (0.161)	Min Abs error ($\log_k$)	0.00039	0.000704
Model	Spherical	Average error ($\log_k$)	0.00076	−0.09371
Lag size (m)	322.3	MAE ($\log_k$)	0.46	0.49
N° of lags	12	MSE ($\log_k$)	0.37	0.43
Neighbors to include	30	RMSE ($\log_k$)	0.61	0.66
-	-	NRMSE ($\log_k$)	0.11	0.15

The target error to be achieved was a global error below 1 order of magnitude and preferably close to 0.5 orders of magnitude. In particular it was assumed that an error between measured and predicted $\log_{10}$k ranging from 0 to 0.4 could represent a proper estimate of k value, while values of $\log_{10}$k error between 0.4 and 0.7 could lead to a reasonable approximation of k values; values over 0.7 and 1.5 were symptoms respectively of bad and very bad estimates. The results were quite satisfactory since the MAE and RMSE were near 0.6 $\log_{10}$k orders of magnitude for both tests.

A reliability map was then created where deviation between measured and estimated values of hydraulic conductivity at different locations was reported. By using the aforementioned classification,

53% of the points (217 pts) were classified as "Good" (abs. error < 0.4), while 22.5% (92 pts) were classified as "Sufficient" (error between 0.4 and 0.7). The remaining 24.5% (100 pts) was classified as "Insufficient" or "Bad", since high errors were observed.

Observation classified as "Bad" are mainly located in the southern part, especially in Chiasso (2 values over 1.5), Stabio (3 pts) and Mendrisio (1 pt) (Supplementary Material: Figure S2). The interpolation reliability here is poor because of the few hydraulic conductivity data available in this aquifer so the results in this location have to be seen considering a greater bias.

The reconstruction of hydrogeological properties was the starting point for the thermal characterization of Quaternary deposits within monitored aquifers. Deposits were divided into fine, medium and coarse litho-textures, according with [29,30] representative values of hydraulic conductivity, corresponding respectively to silt, sand and gravel. Hydraulic conductivity values between 1 and 10^{-3} m/s were considered as "gravel", k between 10^{-3} and 10^{-4} m/s was considered as "sand" deposits and all other deposits showing k values below 10^{-5} m/s were considered as "silt". Through this approach 58% of the aquifers were classified as gravelly, while the remaining portion was classified as sands (39%) and clays (3%).

The classified unconsolidated deposits were thermally characterized by assigning to each lithological class a referential λ value for both a completely wet and completely dry scenario (λ wet and λ dry). This approach was adopted since there are zones located mainly in the northern part of the study area (north of Magadino Plain) and in alluvial fans that are characterized by great groundwater depth. The weighted λ for each pixel was spatially estimated to be:

$$\text{weighted } \lambda = \frac{[(\lambda \text{ dry} * \text{unsaturated thickness} + \lambda \text{ wet} * \text{saturated thickness}]}{100 \text{ m}} \quad (5)$$

A reference bottom elevation for the aquifer of 100 m below ground surface was adopted since no specific data regarding bedrock depth was available. A thermal conductivity of 2 W/mK was assigned to generic Quaternary deposits not belonging to monitored aquifers; it is a reasonable value given the low groundwater depth and the coarse-grained lithologies observed within the subsurface of Cantone Ticino (Table 5). Gravels were assigned a dry thermal conductivity of 0.4 W/mK, 1.7 W/mK for completely saturated medium. Sands and silts were assigned respectively a dry/wet thermal conductivity of 0.5/2.3 W/mK and 0.6/1.4 W/mK.

The results of thermal conductivity reconstruction for both outcrops and sediments are reported in Figure 4. Only one TRT data within unconsolidated material was available for the validation (Supplementary Material: Table S1). The TRT, located near Mendrisio (Laveggio aquifer), returned a λ value of 2.2 W/mK, while the mapped λ reconstruction returned a value of 1.7 W/mK in the same location. The error was therefore quantified in 27% which could be due to the convective contribution of groundwater flow, neglected in this study. A more detailed thermal characterization of the underground through real data would improve the map modeling and the final results.

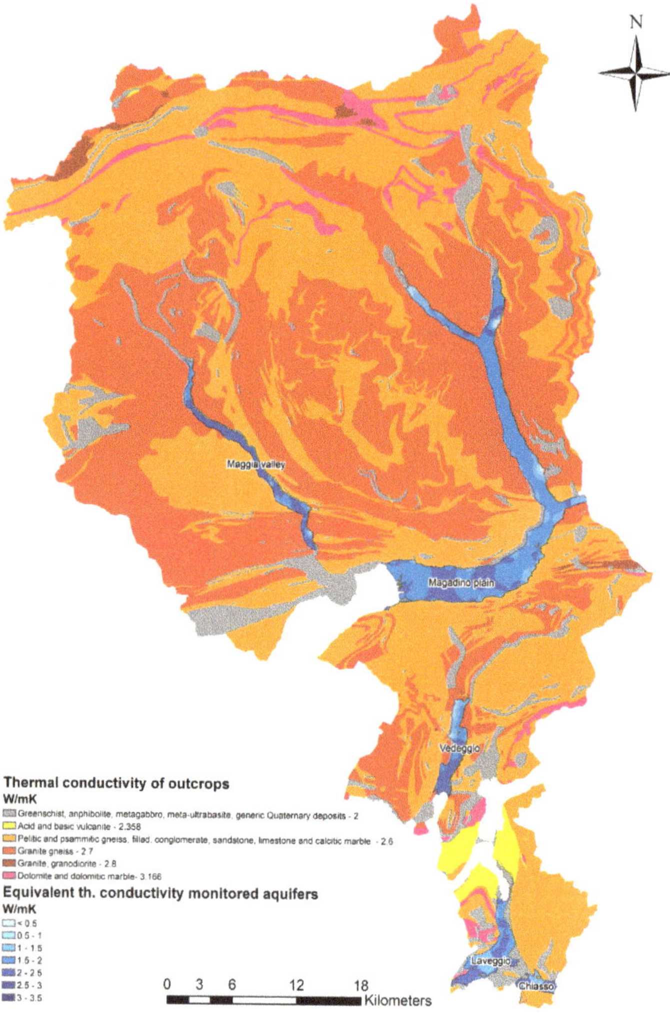

Figure 4. Spatial reconstruction of Cantone Ticino thermal conductivity: red shades belong to outcrops of different lithologies, while blue shades represent monitored aquifers. The names of the monitored aquifers are reported in the map.

3. Technological Constraints

3.1. Target GSHP System Characteristics

The estimation of BHE length is strictly connected to the identification of a reference residential unit that could represent a target building. A residential unit composed of 5 flats of 100 m^2 each (500 m^2 total) that represents the average surface of a flat located in Cantone Ticino was chosen as referential. The multi-family unit was chosen because it is the most common residential type of building observed according to [31]; moreover the Cantonal energy master plan is designed considering a multi-family residential building whose heating is provided by heating oil [32]. The referential GSHP system has been identified in a four probe installation, with a heating capacity of 25 kW, as previously described in Section 1.1.

3.2. Hypothesized Energy Demand

An energetic index of 60 kWh/m^2·year was used as referential to estimate both the annual heat demand and the hot sanitary water (HSW) demand. A MINERGIE® energy standard for refurbished buildings and not for new ones (38 kWh/m^2·year) was chosen in order to stay conservative and to avoid the underestimate of the obtained output BHE length.

MINERGIE® is a quality label used both for new and refurbished buildings that promotes a rational use of energy, the use of renewables and emission savings. This quality label is required for Cantonal or Confederation new buildings. The chosen MINERGIE® index represents the energy demand for buildings that were built before 2000 and are planned to be refurbished.

The product of the MINERGIE® index (60 kWh/m^2·year) for the considered reference area (500 m^2) resulted in 30 MWh/year including domestic hot water: the complete estimated monthly energy profile is reported in Table 7. A summer cooling demand was not taken into account mainly because it could reinstate the thermal field, resulting in COP improvements and reduced simulated BHE length. This is justified since the climate classification for Cantone Ticino is Cfb (wet temperate climate with average summer temperatures <22 °C) therefore the cooling demand is low compared with the heating one, having long winter period and brief summer period.

Table 7. Referential geothermal system parameters and values used in earth energy designer (EED) simulations.

Input	Parameter	Value
Heat exchanger and perforation	Probe type Configuration Distance Borehole diameter Grouting thermal conductivity Volumetric flow rate Probe external diameter Thickness Probes thermal conductivity Distance between internal tubes	Double-U tube 2 × 2 (N° 233)—4 probes 8 m 130 mm 1 W/mK 2 l/s per probe 32 mm 3 mm 0.420 W/mK 80 mm
Thermo-vector fluid	Thermal conductivity Specific heat Density Viscosity Freezing point	0.48 W/mK 3795 J/kgK 1052 kg/m3 0.0052 kg/ms −14 °C
Probes thermal resistance	Constant values - Between fluid and ground - Internal Takes into account internal heat fluxes	 0.140 mK/W 0.450 mK/W
Heating capacity	-	25 kW
Performance Energetic demand	SPF Annual heating demand (comprises HSW)	4 30 MWh 22.5 MWh/year provided by ground 7.5 MWh/year provided by heat pump

	Jan.	Feb.	Mar.	Apr.	May	Jun.	Jul.	Aug.	Sept.	Oct.	Nov.	Dec.
Heating demand [MWh/month]	4.65	4.44	3.75	2.97	1.92	0	0	0	1.83	2.61	3.51	4.32
Peak monthly heating duration [hours/month]	6	5	5	3	3	0	0	0	3	3	5	6

4. Results

4.1. Required BHE Length Calculation

EED software was developed by Lund University for the modeling of vertical BHEs. In very large and complex projects, EED allows for retrieving the optimized required borehole length and layout before starting more detailed analyses. EED is based on g-functions, which depend on the spacing between the boreholes and on the borehole depth. In this paper EED was used to simulate 128 scenarios using a fixed set of system parameters that could represent a referential GSHP system (Table 7) where alternately GST, λ, heat flux and volumetric heat capacity were changed (Supplementary Material Table S3). The range of the values for each parameter partially represents the variability of each value within the mapped areas, but only the most frequent values observed in the most populated zones were taken into account. GST values from 3 to 10 °C were not considered in the simulations because the majority of the closed-loop systems in Ticino are located in zones where GST lies between 11 and 14 °C, therefore adding simulations for these locations would have been redundant. The choice of the intervals was also made in order to speed up computational times, avoiding the realization of thousands of simulations that would have not significantly contributed to the identification of the regression polynomial function. This described procedure allowed creating a good number of scenarios resulting in BHE length values that could be put into relation with GST and λ values (which were previously mapped in Cap. 2). 3D correlation between thermal conductivity, GST and required BHE length was better described by a 2nd grade polynomial function which produces a R^2 of 0.95 and a RMSE of 16 m (Supplementary Material: Figure S3 and Table S2):

$$\text{Required BHE length} = 1236 - 269.1*\lambda - 34.83*\text{GST} + 34*\lambda^2 + 2.47*\lambda*\text{GST} \quad (6)$$

This polynomial regression was applied to GST and λ maps in order to obtain a continuous estimate of BHE length for the previously defined reference building. The resulting map of required BHE length (Figure 5) shows that the potential is lower where the altitude is higher because it is strongly affected by the annual temperature of the underground. A higher potential is mainly located within the major valleys of Cantone Ticino, due to higher annual temperatures and due to the presence of groundwater at low depth. These areas are also affected by restrictions due to the presence of groundwater protection zones, which do not allow a complete deployment of new GSHP systems. The white spots represent the large number of lakes located within Cantone Ticino: there the potential was not mapped.

To understand the reliability of this map the estimated BHE length of the map was compared with the real installed length observed in Ticino for 51 verified systems where heating demand is prevalent. These GSHPs are almost completely located in a portion of territory below 1000 m of altitude, but also four systems installed above 1000 m were observed. To perform this comparison, standardization for the installation power and for the number of probes was executed using the following proportion:

$$\text{Real length} = \frac{\text{estimated required BHE length} * \text{real system power}}{25\ \text{kW}} \quad (7)$$

The BHE length estimated by the map has been corrected for the installed power. This proportion is possible because a linear correlation between cumulative installed BHE length and system power was observed within Cantone Ticino: the resulting map represents the amount of meters needed to obtain 1 kW of installed heating power (m/kW map). The comparison between measured/stated values of m/kW and estimated ones, reported in Figure 6a and Table 8, show that installations with an estimated error ≤10% (27.4% of the total, 14 systems) are located homogeneously within the study area, from North to South and from West to East. 20 systems show an error between 10 and 25% (39.2% of the total), 14 systems (27.4% of the total) show a % error between 25 and 50, while the remaining three systems show values above 50% of error (6% of the total).

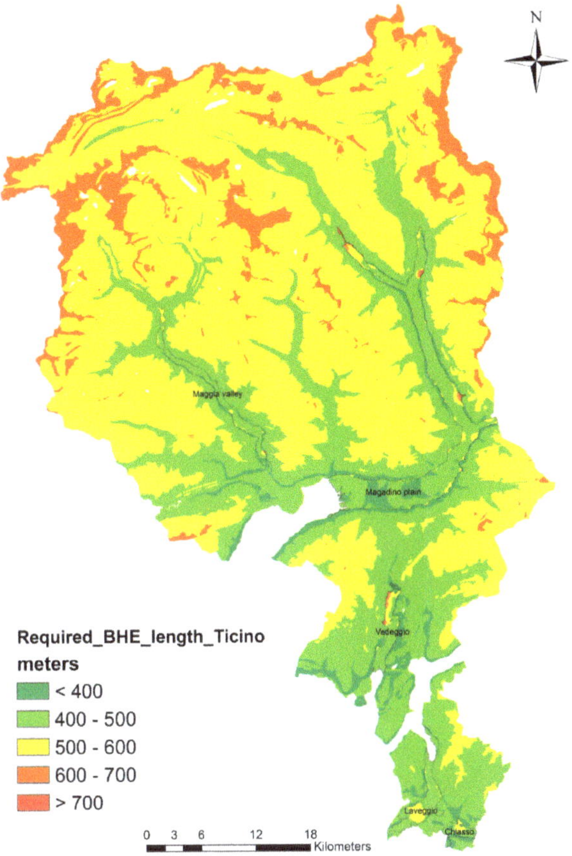

Figure 5. Map of required BHE length to satisfy the imposed energy demand of the referential refurbished MINERGIE® building considered in the simulations.

Table 8. Comparison statistics between measured and computed borehole heat exchanger (BHE) lengths.

Statistic	Value
Max abs error	52 m/kW
Min abs error	1 m/kW
Average error	3.7 m/kW
Mean abs error	6 m/kW
Abs average error (%)	±23%

Basically the same results were also obtained for systems above 1000 m of altitude: the method can be therefore reliable for different ranges of elevation. Results globally show that the created map has a semi-quantitative value at regional scale, with an overall average error of 23% equivalent to ±6 m/kW. Considering that the test area is approximately 3000 km^2 wide, results are satisfactory. The less performant aquifer results to be Laveggio, which requires on average 18.4 m to provide 1 kW of installed capacity, while the most performant aquifer is Chiasso, requiring on average 17.4 m/kW. These values must be considered by taking into account the incertitude of the estimated m/kW map, which is ±23%. Starting from this premise the estimated aquifers potential results to be quite similar: a general uniformity of geo-exchange potential can be hypothesized.

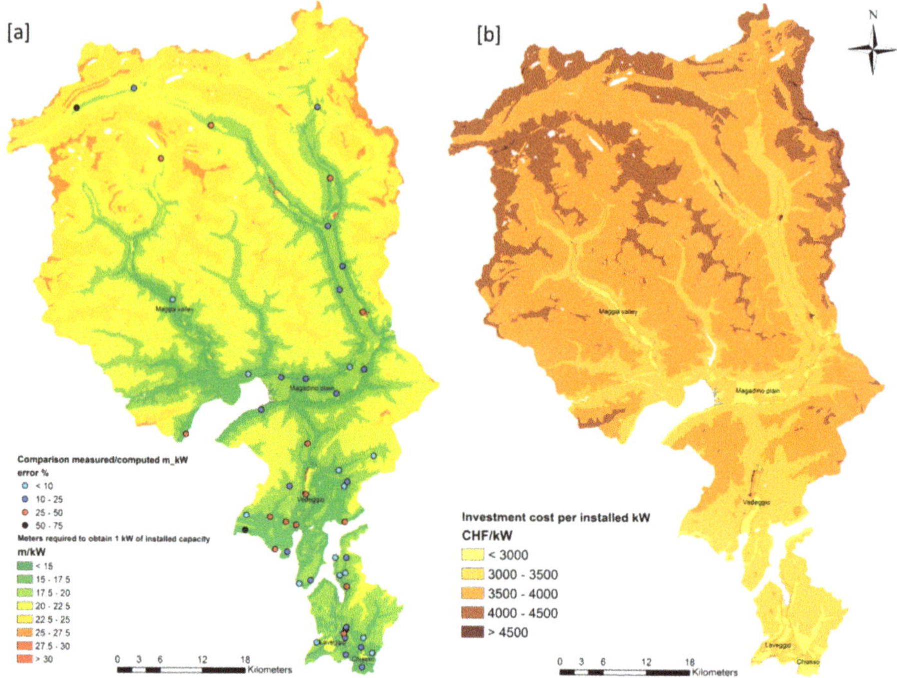

Figure 6. Map of (**a**) estimated meters required to obtain 1 kW of installed power and (**b**) specific capital cost expressed as CHF/kW.

4.2. Economical Maps

4.2.1. Total Installation Costs and CHF/Kw

Another interesting topic was related to the estimation of an overall cost of the hypothesized GSHP referential system depending on its length. Different drillability values to each macro-lithology observed within Cantone Ticino and reported in Table 9 were assigned. These costs data came from an estimate made by SUPSI in collaboration with a drilling company. Drilling costs are approximately quantified between a minimum of 60% of the total installation costs for consolidated sedimentary rocks and a maximum of 84% for hard rocks. Drilling into hard volcanic rocks would cost more than drilling into sediments or sedimentary rocks. Probe cost and grouting costs refer to the installation of a double-U probe, which is the most common probe type used in Cantone Ticino. Labor costs are approximately quantified in 4% of the total cost while the disposal of the drilling mud in a controlled landfill (mandatory for Switzerland) represents 7% of the total cost. All these information plus estimates of labor costs and mud disposal were implemented in GIS environment and an installation costs map was produced (Figure 6b). Cost estimates have to taking into account an average error of ±23% in m/kW evaluation. The final cost estimate including labor costs and mud disposal ranges from 70,000 CHF to 133,000 CHF for the identified referential residential system. At the moment of writing, 1 CHF or Swiss Franc is equal to 0.87€. This means that each apartment/family should spend between 14,000 ± 3200 CHF and 27,000 ± 6200 CHF (approx. between 12,000 ± 2760 and 23,500 ± 5400€) for the installation of the geothermal system, depending on location and on-site potential. The total costs were then divided by the installed heat capacity (25 kW) to calculate a specific capital cost (CHF/kW). An advantageous cost per installed kW of a GSHP system is estimated to be between 3000 and 3500 CHF/kW (2600–3000 €/kW). A potential user/stakeholder could also invest more money, approximately between 3500 and 4000 CHF/kW (3000–3500 €/kW), but the investment would become less and less profitable with longer payback periods and loss of attractiveness.

Table 9. Economic parameters for the costs estimate: the drilling technique used for consolidated sedimentary rocks and unconsolidated material was direct flush mud rotary while the drilling technique used within hard rock is down the hole-hammer.

Parameters	Hard Rock	Consolidated Sedimentary Rock	Unconsolidated Material
Drillability (CHF/m)	91.2	105.2	101.5
Heat Pump: cost per kW (CHF/kW) *		1000	
Probe cost (CHF/m) **		15.7	
Grouting (CHF/m) ***		6.6	

* Estimated within Cheap-GSHPs based on a GALLETTI ENE009 heat pump model; ** REHAU—RAUGEO Probe PE-Xa green; *** REHAU—Raugeo fill red. Heat pump, probe and grout cost the same independently from ground type.

4.2.2. MATLAB/Octave Tool for Payback Period Estimate

A crucial topic was to assess the potential economic attractiveness of closed-loop shallow geothermal energy systems against the current heating systems in Cantone Ticino, which are mainly fossil ones (heating oil and natural gas, almost completely methane). The challenge was to spatialize this information, putting it on map. Global cost of the geothermal system were evaluated by mainly considering drilling depth and heat pump cost, plus the evaluation of maintenance and ancillary costs. The geothermal costs were compared with costs for fossil heating systems through the estimation of cumulative cash flows (savings and expenses) through an experimental MATLAB/Octave [33,34] tool that could estimate a series of economic/environmental indicators. The economic indicators are the payback period (the time after which the investment starts being profitable), the Net Present Value (NPV) of the investment, which is an economic index used to preliminarily evaluate the profitability of an investment before actually doing it, determining the current value of all future cash flows generated by a project after accounting for the initial capital investment. The tool also calculates the amount of saved CO_2 emissions against referential oil and natural gas systems. This tool was initially developed for the purpose of mapping procedure, but it is a stand-alone application that can be used by any type of user which would like to have a preliminary economic feasibility of its geothermal project against one relying on fossil fuels. The complete description of the tool is reported in [Supplementary Material, Economic tool]. The user provides the tool with a series of information concerning a closed-loop geothermal system and concerning the fossil fuel system (heating oil or natural gas) that he wants to use as comparison. The tool calculates the costs for each system and the savings between a geothermal system and the chosen fossil fuel one. The tool also returns a payback period: by performing a series of simulations a linear relationship between BHE length and payback period for both heating oil and natural gas was found, considering a period of 25 years for the calculation of the cash flow.

The empirical relations are:

$$\text{Payback period heating oil (years) without economic subsidies} = (0.0274 * \text{required BHE length}) - 4.2941 \quad (8)$$

$$\text{Payback period natural gas (years)} = (0.0273 * \text{required BHE length}) - 3.1912 \quad (9)$$

Having a BHE length map and this relations allowed building experimental payback period maps, which are reported in Figure 7a for heating oil and in Figure 7b for natural gas.

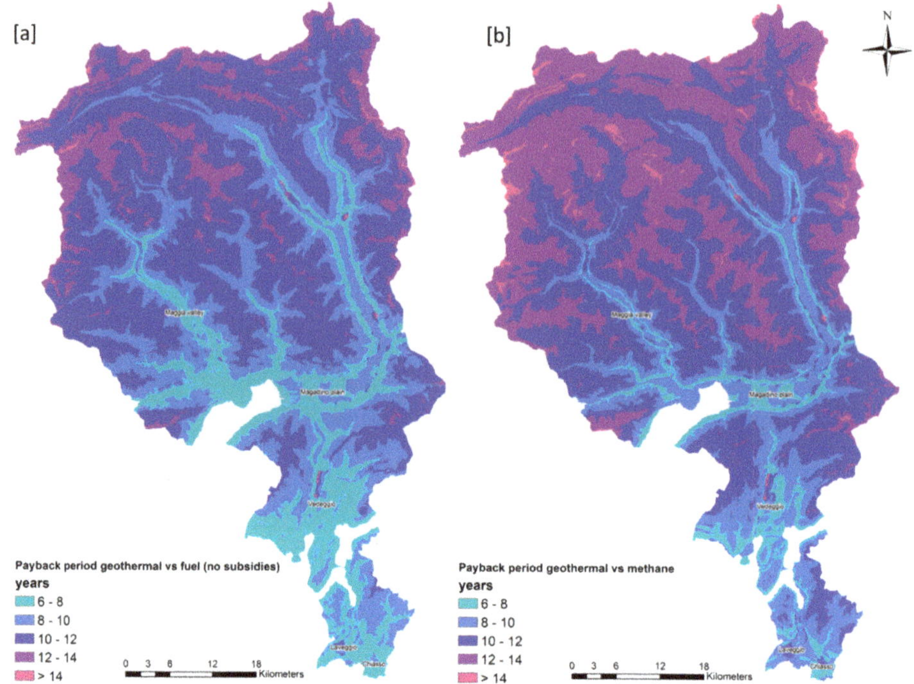

Figure 7. Maps of estimated payback period resulting from a geothermal vs heating oil analysis without considering economic subsidies (**a**) and geothermal vs natural gas (**b**).

The maps show that the payback period time for the considered referential geothermal system mainly ranges from 7 to 15 years, depending on thermal characteristics of the underground and its temperature. Payback periods $\geq$ 14years are only observed at elevations >2000 m, where the very low population density renders this assessment largely irrelevant, or in alluvial fans.

As previously said, aquifers host a greater potential for the installation of GSHP systems, mainly due to their presence at lower elevations and due to their larger thermal conductivities, caused by low groundwater depth (except for alluvial fans). On average the most promising place to install closed systems would be Chiasso aquifer, which requires on average approximately 7 years and 6 months to payback the investment when replacing an old heating oil system, while the less attractive aquifer is Laveggio requiring approximately 8 years and 6 months to payback the investment. The presence of Cantonal subsidies for replacing the oil system with a new geothermal one would cut down the payback periods of approximately one year.

The payback period estimate for a natural gas system (Figure 7b) states that it will take more time to return from the initial investment, (usually one year more, at least), since a natural gas system has much lower installation costs and similar operating costs than a heating oil one. The mapped values of payback periods must be seen considering an average error of $\pm$23% within the calculated m/kW maps, but the produced maps still can highlight regions with longer or shorter payback periods.

The last created map was related to the profitability of the investment calculated from NPV. This relation was estimated by running several simulations using the economic tool for both oil and natural gas systems, considering an economic analysis of 25 years and a discount rate of 3% (the return that could be earned per unit of time on an investment with similar risk). Results showed that NPV turned negative for both comparisons when the required BHE length depth was >500 m. Figure 8 reports this information by classifying the study area into "profitable" and "unprofitable" depending on required BHE respectively lower or higher than 500 m. When NPV is >0 the investment is profitable (it adds monetary value to the capital), otherwise it is unprofitable (does not add monetary value, the user should invest in some other projects). As expected, the map states that the investment in a new geothermal system compared to oil/methane system would be profitable in all valleys and aquifers and even at elevations of approximately 600 m a.s.l.

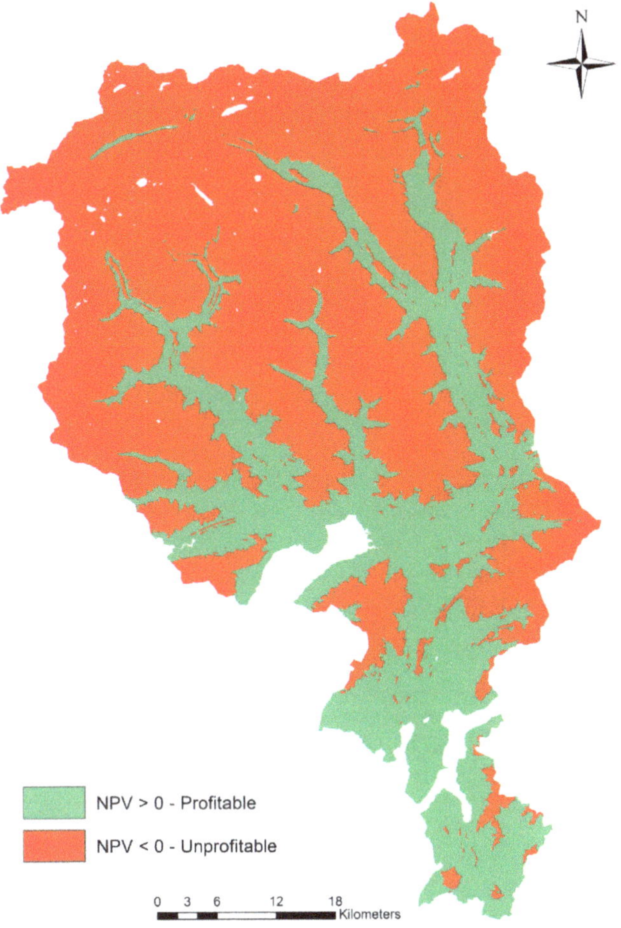

Figure 8. Map of profitability of the investment calculated for an economic analysis of 25 years. The areas where the investment would be profitable considering a discount rate of 3% (the return that could be earned per unit of time on an investment with similar risk) are reported in green. Red portion are zones where the investment in a closed-loop system would not be profitable, considering the NPV.

4.3. CO_2 Savings

Swiss electrical energy is a type of low emission electric energy, because it is basically produced without fossil energetic vectors [35]. At the other hand the consumed electricity by the users has higher CO_2 emissions, since the amount of CO_2 emissions from the imported energy also have to be considered. Values of t/CO_2 emissions per year contained in [35] were considered as referential and used in the estimation of total carbon dioxide emissions.

The economic tool contains a part devoted to the calculation of the CO_2 emissions given different type of energy vector. It estimates the potential savings in CO_2 emission by the adoption of a closed-loop geothermal system instead of heating oil or natural gas one. This is explained more in detail in [Supplementary Material, Economic tool]. Comparing the geothermal solution against fossil fuel systems and considering a COP of 4 (used in all EED simulations), the amount of saved CO_2 emissions would be 6.8 tCO_2/year for heating oil and 4 tCO_2/year for natural gas. Considering a COP from 3 (worst case scenario) to 5 (best scenario) the emission savings would range from 6.3 to 7.1 tCO_2/year for heating oil and from 3.5 to 4.3 tCO_2/year for natural gas. The calculation of the savings of CO_2 emissions is conceptually useful only if the geothermal system is chosen as an alternative/replacement to current heating systems. New systems do not provide any emission reduction and the geothermal systems that replace fossil fuel ones must also operate with electricity coming from low to zero CO_2 emissions, as happens for Switzerland.

4.4. Comparison of Produced Techno-Economic Maps with Available Regulation Maps

A comparison between m/kW map and the map related to closed-loop systems regulation is reported in Figure 9. The figure shows four different locations that were chosen within Cantone Ticino: all images compare Au areas where new closed-loop systems are allowed to be installed (bordered in blue) with the estimated potential in form of m/kW map. In the second and fourth frame (upper right and lower right of Figure 9), two zones of the middle and southern part of Cantone Ticino are reported. The upper part of Vedeggio aquifer hosts a very low geothermal potential, mainly due to higher groundwater depth given the presence of many alluvial fans. The geothermal potential is observed to be very low, with consequent higher installation costs and operating costs due to deeper drilling and lower geo-exchange. The peculiarity of Au conflict zones is also that they are mainly located within alluvial fans or zones with high groundwater depth, with a consequently lower geothermal potential. The crucial aspect is that the areas where the installation of vertical geothermal probes is allowed often show a lower techno-economic potential. This means that the majority of BHEs are located in areas of low potential. At the other hand, results show that zones with higher techno-economic potential are interested by the occurrence of groundwater protection zones, where the installation of closed-loop systems is not allowed. Considering these results, it is clearly observable how the actual allocation of thermal resources reflects on the performances and costs of the systems. Given all the uncertainties that are intrinsic to a regional mapping as the proposed one, the allocation seems to be unoptimized, clearly raising regulation improvements.

Thermal interference and thermal resource management will be some of the most important topics to face in the next years: this could be assessed with more empirically verified maps.

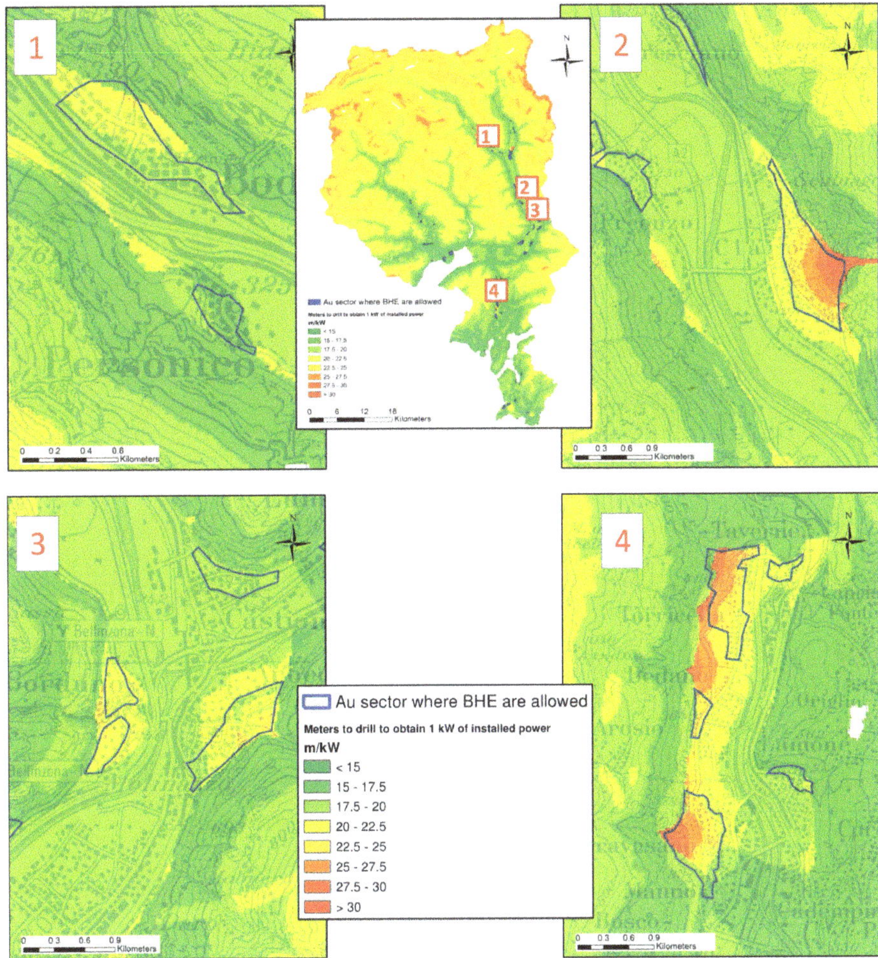

Figure 9. Comparison in some test areas between shallow geothermal potential expressed as m/kW and areas where the installation of new BHE is allowed.

5. Discussion

Results showed that the referential building/GSHP system would require a BHE length between 400 and 700 m or, if expressed as an index, between 16 and 28 m for every kW of installed heating capacity (m/kW). Lower values of this index (and therefore a higher potential) are found within major valleys: this happens because the mean ground temperature is higher during the year and due to the shallower depth of the groundwater table, resulting in higher values of sediments thermal conductivity. m/kW map was compared against 51 real geothermal systems and the average error was quantified in 23% or ±6 m/kW. Further studies could focus on more detailed hydrogeological and thermal characterization of the subsurface using stored well logs and the detailed assessment of bedrock depth for each aquifer.

A spatial economic estimate of system costs was performed using documented prices and installation costs: it shows that this type of referential system could cost approximately between 80,000 CHF and 110,000 CHF (70,000 and 96,000€), with an estimated specific capital cost index between 3000 and 4500 CHF (2600 and 3900€) per installed kW (CHF/kW).

The estimated payback period for heating oil and natural gas systems ranges from approximately 7 to 15 years, depending on ground characteristics and air temperature. Regardless the intrinsic incertitude of the mapping process, natural gas systems require more time to be paid back compared to a heating oil system, because their capital cost is much lower than a heating oil system.

Sectors where new GSHP systems are allowed are often the locations where a lower geothermal potential is observed. This results in a non-optimized management of the geothermal resource that can lead to costs increase, efficiency decrease and potential loss of attractiveness of this renewable energy.

Moreover, in the next years, a strong interference issue between adjacent systems will probably occur, since closed-loop systems are only allowed in small portions of territory and a stable amount of new requests is observed. In light of these discoveries, a review of the current Cantone Ticino regulations regarding shallow geothermal energy and a more optimized allocation of the thermal resource should be strongly advised. Below some potential solutions (which do not consider the techno-economic-political feasibility):

- *Widening of allowable areas*: would promote the installation in new areas, but this would result in an increase of new requests, posing a serious threat to groundwater quality (from a chemical and thermal standpoint);
- *Creation of large shallow geothermal systems that could do thermal storage plus district-heating*: by creating new shallow geothermal installations in areas with high geo-exchange potential and by delivering the produced heat to areas having lower potential could optimize the management of the resource, since less geothermal systems in low potential areas would be needed. This solution could imply the use of Borehole Thermal Energy Storage systems, constituted of a large number of geothermal probes with short mutual spacing, which is beneficial in order to avoid leakage of heat/"cold" outside the borehole array [36].

Given the complexity of the input parameters and of the procedure, these mapping results are a good starting point for further analysis and characterization of the subsurface, in order to obtain more accurate maps. This could be done in the future through a systematic collection and publication by regulatory agencies of subsurface materials properties, through the development of an accessible database. In this way, map products would enhance regulatory efforts over time, as well as assist entities involved in the design and economic analyses of GSHP systems.

Supplementary Materials: The following are available online at http://www.mdpi.com/1996-1073/12/2/279/s1, Figure S1: MAAT map of Cantone Ticino with chosen air temperature MeteoSwiss stations, Table S1: Measured thermal conductivity from TRT executed within Cantone Ticino, Figure S2: Reliability of litho-textural characterization of the subsurface, Table S2: 3D regression coefficients and fit for the identification of the BHE length regression, Figure S3: 3D regression between thermal conductivity (x1), Ground Surface Temperature (x2) and estimated BHE length (y), Table S3: Summary of all the 128 EED simulations performed to identify the polynomial regression between λ, GST, BHE length.

Author Contributions: R.P. wrote the paper, performed GIS calculations and mapping, plus EED simulations. R.P., S.P. and A.G. conceived the paper and the method used to create maps of shallow geothermal potential. S.P. and A.G. reviewed the paper and provided suggestions for its improvement. Sebastian Pera and Antonio Galgaro provided funding for the execution of the work and for its publication.

Funding: The work is related to Cheap-GSHPs project, which has received funding from the European Union's Horizon 2020 research and innovation programme under grant agreement No. 657982. The project aims at reducing the installation costs while improving the efficiency of closed-loop GSHP systems through the development of new technologies. SUPSI was involved in the realization of multi-parameter geothermal maps regarding the closed-loop shallow geothermal potential, therefore a methodology had to be found: the design and testing of the methodology was performed hereafter on Cantone Ticino. The work performed by SUPSI within Cheap-GSHPs was funded by the Swiss State Secretariat for Education, Research and Innovation (SERI) under contract No. 15.0176.

Acknowledgments: The authors would like to thank Michele Fre of Luzi Bohr-Drilling AG for the estimate of drilling costs for different lithologies, and Marco Belliardi from SUPSI for the support in doing EED simulations. The authors thank Galletti (Belgium) for providing them with some referential costs of their heat pumps and REHAU (Germany) for providing referential costs of grouting material. The authors would also like to thank the two anonymous reviewers that improved the paper through their suggestions and comments; moreover the authors thank the academic editor for its hints and insights.

Conflicts of Interest: The authors declare no conflict of interest.

Abbreviations and Symbols

SUPSI	Scuola Universitaria Professionale della Svizzera Italiana
GSHP	Ground Source Heat Pumps
COP/EER	Coefficient of Performance/Energy Efficiency Ratio
MAAT/GST	Mean Annual Air Temperature/Ground Surface Temperature
MAE	Mean Absolute Error
RMSE	Root-Mean-Square Error
k	Hydraulic conductivity
λ	Thermal conductivity
TRT	Thermal Response Test
BHE	Borehole Heat Exchanger
EED	Earth Energy Designer
DHW	Domestic Hot Water
SPF	Seasonal Performance Factor
CHF	Swiss franc
NPV	Net Present Value

References

1. Lund, J.W.; Boyd, T.L. Direct utilization of geothermal energy 2015 worldwide review. *Geothermics* **2016**, *60*, 66–93. [CrossRef]
2. Swiss Federal Office of Energy SFOE. Available online: http://www.bfe.admin.ch/index.html?lang=en (accessed on 10 January 2019).
3. MINERGIE®. Available online: https://www.minergie.ch/fr/certifier/minergie/?l (accessed on 10 January 2019).
4. Perego, R.; Guandalini, R.; Fumagalli, L.; Aghib, F.S.; de Biase, L.; Bonomi, T. Sustainability evaluation of a medium scale GSHP system in a layered alluvial setting using 3D modeling suite. *Geothermics* **2016**, *59*, 14–26. [CrossRef]
5. Consiglio di Stato Della Repubblica e Cantone Ticino, Regolamento Sull'utilizzazione Dell'energia (RUEn). 2008. (In Italian)Available online: https://www3.ti.ch/CAN/RLeggi/public/index.php/raccolta-leggi/legge/vid/09_36 (accessed on 10 January 2019).
6. Overview on Cantone Ticino's Energy, Cantonal Office of Statistics, February 2018. Available online: https://www3.ti.ch/DFE/DR/USTAT/allegati/prodima/4308_energia.pdf (accessed on 10 January 2019).
7. Environment Division, Office of water Protection and Water Supply, State Council Proceedings 2016–Statistical Annex (Public Report). Available online: https://m4.ti.ch/fileadmin/CAN/TEMI/RENDICONTOCDS/2016/RENDICONTO/Allegato_statistico_2016_documento_completo.pdf (accessed on 10 January 2019).
8. GESPOS. Available online: https://geoservice.ist.supsi.ch/gespos/ (accessed on 10 January 2019).
9. Swiss Waters Protection Ordinance (WPO). 2010. Available online: https://www.admin.ch/opc/en/classified-compilation/19983281/index.html (accessed on 10 January 2019).
10. Swiss Federal Office for the Environment FOEN, Exploitation de la Chaleur Tirée du Sol et du Sous-Sol. 2009. Available online: https://www.bafu.admin.ch/bafu/fr/home/themes/eaux/publications/publications-eaux/exploitation-chaleur-tiree-sol-sous-sol.html (accessed on 10 January 2019).
11. Thüring, M. Regolamentazione Dello Sfruttamento Dell'energia Geotermica nel Canton Ticino, Sezione per la Protezione Dell'aria, Dell'acqua e del Suolo, Canton Ticino. 2006; (internal report).
12. Lopez, A.; Roberts, B.; Heimiller, D.; Blair, N.; Porro, G. *US Renewable Energy Technical Potentials: A GIS-Based Analysis*; United States National Renewable Energy Laboratory (NREL): Golden, CO, USA, 2012.
13. Schiel, K.; Baume, O.; Caruso, G.; Leopold, U. GIS-based modelling of shallow geothermal energy potential for CO_2 emission mitigation in urban areas. *Renew. Energy* **2016**, *86*, 1023–1036. [CrossRef]
14. García-Gil, A.; Vázquez-Suñe, E.; Alcaraz, M.M.; Juan, A.S.; Sánchez-Navarro, J.Á.; Montlleó, M.; Rodríguez, G.; Lao, J. GIS-supported mapping of low-temperature geothermal potential taking groundwater flow into account. *Renew. Energy* **2015**, *77*, 268–278. [CrossRef]

15. Bertermann, D.; Klug, H.; Morper-Busch, L. A pan-European planning basis for estimating the very shallow geothermal energy potentials. *Renew. Energy* **2015**, *75*, 335–347. [CrossRef]
16. Gemelli, A.; Mancini, A.; Longhi, S. GIS-based energy-economic model of low temperature geothermal resources: A case study in the Italian Marche region. *Renew. Energy* **2011**, *36*, 2474–2483. [CrossRef]
17. Casasso, A.; Sethi, R. Assessment and mapping of the shallow geothermal potential in the province of Cuneo (Piedmont, NW Italy). *Renew. Energy* **2017**, *102*, 306–315. [CrossRef]
18. Arola, T.; Eskola, L.; Hellen, J.; Korkka-Niemi, K. Mapping the low enthalpy geothermal potential of shallow Quaternary aquifers in Finland. *Geotherm. Energy* **2014**, *2*, 9. [CrossRef]
19. Viesi, D.; Galgaro, A.; Visintainer, P.; Crema, L. GIS-supported evaluation and mapping of the geo-exchange potential for vertical closed-loop systems in an Alpine valley, the case study of Adige Valley (Italy). *Geothermics* **2018**, *71*, 70–87. [CrossRef]
20. Galgaro, A.; di Sipio, E.; Teza, G.; Destro, E.; de Carli, M.; Chiesa, S.; Zarrella, A.; Emmi, G.; Manzella, A. Empirical modeling of maps of geo-exchange potential for shallow geothermal energy at regional scale. *Geothermics* **2015**, *57*, 173–184. [CrossRef]
21. Hellström, G.; Sanner, B. Earth Energy Designer 3.2. Manual. Available online: https://www.buildingphysics.com/manuals/EED3.pdf (accessed on 10 January 2019).
22. SIA Schweizerischer Ingenieur- und Architekten-Verein. *Sondes Géothermiques*, 1st ed.; SIA: Zurich, Switzerland, 2010.
23. MeteoSwiss. Available online: https://www.meteoswiss.admin.ch/ (accessed on 10 January 2019).
24. Signorelli, S.; Kohl, T. Regional ground surface temperature mapping from meteorological data. *Glob. Planet. Chang.* **2004**, *40*, 267–284. [CrossRef]
25. IDAWEB. Available online: https://gate.meteoswiss.ch/idaweb/login.do (accessed on 10 January 2019).
26. Federal Office of Topography Swisstopo. Available online: https://www.swisstopo.admin.ch/ (accessed on 10 January 2019).
27. Linda, S. Interplay between Opposite Vergence Thrusts along the Southern Alps Margin in Canton Ticino (Switzerland): Geometry and Kinematics in Support of the Characterization of Geothermal Potential. Ph.D. Thesis, Pavia, Italy, 2015.
28. Luo, J.; Tuo, J.; Huang, W.; Zhu, Y.; Jiao, Y.; Xiang, W.; Rohn, J. Influence of groundwater levels on effective thermal conductivity of the ground and heat transfer rate of borehole heat exchangers. *Appl. Therm. Eng.* **2018**, *128*, 508–516. [CrossRef]
29. Freeze, R.A.; Cherry, J.A. *Groundwater*; Prentice-Hall: Englewood, NJ, USA, 1979.
30. Fetter, C.W. *Applied Hydrogeology*, 4th ed.; Prentice Hall: Upper Saddle River, NJ, USA, 2001.
31. Federal Office of Statistics, Statistics of Buildings and Residential Buildings, Neuchâtel. Available online: http://www3.ti.ch/DFE/DR/USTAT/allegati/prodima/4509_costruzioni_e_abitazioni.pdf (accessed on 10 January 2019).
32. Republic and Cantone Ticino–Cantonal Energetic Plan. Available online: http://www4.ti.ch/generale/piano-energetico-cantonale/tema/tema/ (accessed on 10 January 2019).
33. MATLAB. Available online: https://www.mathworks.com/products/matlab.html (accessed on 10 January 2019).
34. GNU OCTAVE. Available online: https://www.gnu.org/software/octave/ (accessed on 10 January 2019).
35. Swiss Federal Office for the Environment FOEN. *Compilation of the CO_2 Emission Factors and Energy Content of the Various Energy Sources that Are Used in the Greenhouse Gas Inventory*; Swiss Federal Office for the Environment FOEN: Ittigen, Switzerland, 2016.
36. Banks, D. *An Introduction to Thermogeology: Ground Source Heating and Cooling*; John Wiley & Sons: Hoboken, NJ, USA, 2012.

Article

Comparative Analysis of Different Methodologies Used to Estimate the Ground Thermal Conductivity in Low Enthalpy Geothermal Systems

Cristina Sáez Blázquez *, Ignacio Martín Nieto, Arturo Farfán Martín, Diego González-Aguilera and Pedro Carrasco García

Department of Cartographic and Land Engineering, University of Salamanca, Higher Polytechnic School of Avila, Hornos Caleros 50, 05003 Avila, Spain; nachomartin@usal.es (I.M.N.); afarfan@usal.es (A.F.M.); daguilera@usal.es (D.G.-A.); retep81@usal.es (P.C.G.)

* Correspondence: u107596@usal.es; Tel.: +34-675-536-991

Received: 3 April 2019; Accepted: 29 April 2019; Published: 2 May 2019

Abstract: In ground source heat pump systems, the thermal properties of the ground, where the well field is planned to be located, are essential for proper geothermal design. In this regard, estimation of ground thermal conductivity has been carried out by the implementation of different techniques and laboratory tests. In this study, several methods to obtain the thermal properties of the ground are applied in order to compare them with the reference thermal response test (TRT). These methods (included in previous research works) are carried out in the same geological environment and on the same borehole, in order to make an accurate comparison. All of them provide a certain value for the thermal conductivity of the borehole. These results are compared to the one obtained from the TRT carried out in the same borehole. The conclusions of this research allow the validation of alternative solutions based on the use of a thermal conductive equipment and the application of geophysics techniques. Seismic prospecting has been proven as a highly recommendable indicator of the thermal conductivity of a borehole column, obtaining rate errors of below 1.5%.

Keywords: ground source heat pump; thermal conductivity; thermal response test; thermal conductive equipment; geophysics

---

1. Introduction

The global growing energy needs have sparked renewed interest in ground source heat pump systems. These systems are traditionally used for space heating and cooling by the extraction of the ground's energy through a borehole heat exchanger (BHE) [1]. High initial investments are commonly associated with these installations so that an optmal ground loop dimensioning is advisable to avoid unnecessary costs. In this regard, the design process requires knowing rather accurately the thermal conductivity of geological formation where the ground source heat pump (GSHP) system will be located [2,3]. Ground thermal conductivity is usually determined by the implementation of a Thermal Response Test (TRT), the main purpose of which is the measuring of the equivalent thermal conductivity of the ground volume tested and the thermal resistance of the BHE [4–7]. The conventional TRT is based on circulating heated water in a closed loop, which simulates heat transfer occurring in a ground heat exchanger. Inlet and outlet water temperatures and flow rate are measured during the test. These data are then analyzed by the implementation of analytical or numerical models that allow determining the ground thermal conductivity and the borehole thermal resistance [8,9]. The most used interpretation technique relies on the first-order approximation of the infinite line-source model. Assuming a constant heating power, a linear regression model is fitted to the late temperature measurements to calculate the mean time derivative of the temperature and deduct the desired parameters. This first approach to linearize the infinite line source model requires rejecting the early measurements for the subsequent test interpretation. TRT duration is usually established between 36-72 h, but this duration has been thoroughly discussed in the past [10] and at the moment is an area of active research [11–15]. Despite the errors these test could involve [4], the high accuracy of ground thermal conductivity results represent essential information in the corresponding geothermal loop sizing. The main inconvenience associated with the realization of a TRT is its relatively high cost (around 3000 Euros), remaining an issue that prevents its widespread use. This problem is especially significant in small installations, where the test increases the initial investment without clear compensation.

Focusing on alternative solutions, some variations of a TRT are available in the existing literature. For example, Henke et al. [16] proposed an experimental apparatus that, as a TRT, measures the temperature response of a borehole. Freifeld et al. [17] developed a borehole methodology to estimate the formation thermal conductivity in situ with spatial resolution of one meter. However, the alternative techniques found in other authors' researches have similar economic issues [18] and their validity is still unclear.

In this research, a series of methodologies (already published and available in the current literature), aimed at the estimation of the ground thermal conductivity, are applied on a real area. These techniques are then evaluated by their comparison with the results of a thermal response test carried out on a borehole placed in the same location. In a nutshell, the thermal conductivity of a certain geological environment is determined by the implementation of affordable methods whose validity is thoroughly assessed. Thus, the main problematic addressed in this work is characterization of ground thermal from different methods to generally improve the design of a low enthalpy geothermal system.

2. Materials and Methods

2.1. Global Description of the Area under Study

The comparison of the methodologies considered in this research is derived from the thermal characterization of a 43 m length and 220 mm diameter borehole placed in the province of Ávila (Spain). The exact location of this well is detailed in Table 1.

Table 1. Location of the borehole where this research is focused.

Borehole Position	
Latitude	40°39′2,45 N
Longitude	4°40′44,84 O

The area contemplated in the present research is located in the center of Ávila (Spain). This region is geologically constituted by two main blocks. One of them is defined by igneous and metamorphic rocks from the Upper Carboniferous-Low Permian and Pre-Cambrian-Low Cambrian periods, respectively. The second block is characterized by the presence of sedimentary materials from the Mesozoic, Tertiary and Quaternary (oriental area of the Amble's valley) periods [19,20]. In the case of the volume of ground considered here, it belongs to granite formations and more specifically, adamellite rocks. This information can be deduced from the geological map of the region presented in Figure 1.

For a more precise geological characterization, geophysical tests were applied on the experimental borehole. A well logging system was used to obtain the specific earth information. It consists of the measuring of continuous and simultaneous record of different physical parameters throughout the borehole column. The equipment used for the mentioned purpose utilizes a series of interchangeable multi-parameter sensors that allowed to register the following parameters: spontaneous potential, resistivity, and natural gamma radiation. Figure 2 includes the register of the well logging test applied on the study borehole and the stratigraphic column derived from its interpretation.

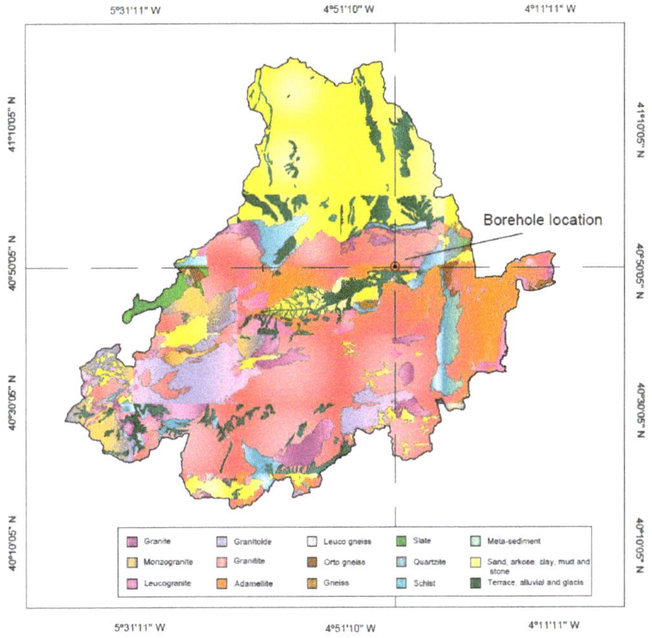

Figure 1. Principal geological formations integrating the region of Ávila [21].

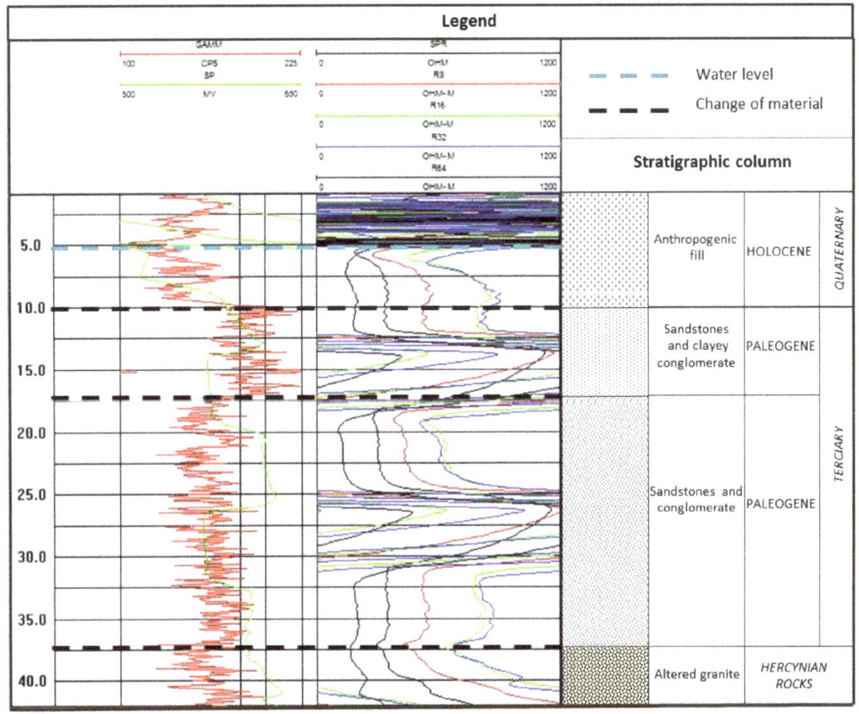

Figure 2. Well logging data and stratigraphic column of the borehole under study.

Once the geological levels that constitute the ground in the area of the borehole were accurately defined, different tests were conducted in the area with the objective of determining the thermal conductivity of the materials previously detected. The implemented methodologies are described in the following subsections.

2.2. Thermal Conductivity Characterization

The principal purpose of this section is the estimation of the thermal conductivity of the borehole column described above. To that end, different methods and procedures implemented in previous author's researches are considered here to finally compare them with the results of a TRT. The following subsections contain the description of each of the mentioned thermal conductivity estimator techniques.

2.2.1. KD2 Pro Measurements

In a previous research work, the thermal conductivity map of the province of Ávila (study area of this research) was created from experimental measurements on the principal geological formations of the region. Representative rocky samples were collected and taken to the laboratory, where the thermal conductivity parameter was measured.

After systematic sample processing—drilling and carving obtaining samples with a specific size, removal of excess material—KD2 Pro equipment was used to measure the thermal conductivity of each geological formation. Before its use, a hole of 6 cm length and 3.9 mm in diameter was made on each rocky sample in order to introduce the RK-1 sensor of KD2 Pro device. More information about the specific measuring methodology is provided in the full published version of the manuscript [21].

As a result of the KD2 Pro measurement, the mentioned research provides the thermal conductivity of the rocky and soil formations. According to the borehole column in Figure 2, the volume of ground under study is constituted by different layers of materials. In order to obtain a representative thermal conductivity value of the whole column well, the thermal conductivity of each layer and its thickness must be considered. Based on the results offered in the research, thermal conductivities of the borehole materials were deduced. All this information is included in Table 2. Thermal conductivity values presented in Table 2 correspond to the average values registered for each geological formation in the manuscript considered here. However, for the last layer of altered adamellites, the lowest values of the mentioned study were selected due to the presence of loose materials and the altered state of granite rocks in that level.

Table 2. Borehole column, geological layers, thicknesses and thermal conductivity values.

	Geological Composition	Thickness (m)	Thermal Conductivity (W/mK) *
Layer 1	Anthropogenic fills	10	1.502
Layer 2	Sandstones and clayey conglomerate	7.5	1.882
Layer 3	Sandstones and conglomerate	20	2.041
Layer 4	Altered adamellite	5.5	2.565

* According to the consulted research [21].

Finally, the thermal conductivity representative of the whole studied borehole can be obtained from the application of Equation (1) and the information previously attached in Table 2.

$$k_T(\mathrm{W/mK}) = k_1 \cdot T_1 + k_2 \cdot T_2 + k_3 \cdot T_3 + k_4 \cdot T_4 \tag{1}$$

where:

k_T = Global thermal conductivity of the whole borehole column.
k_1 = Thermal conductivity of the geological formation of layer 1.
k_2 = Thermal conductivity of the geological formation of layer 2.
k_3 = Thermal conductivity of the geological formation of layer 3.

k_4 = Thermal conductivity of the geological formation of layer 4.
T_1 = Thickness of layer 1 expressed as a percentage of the total well thickness.
T_2 = Thickness of layer 2 expressed as a percentage of the total well thickness.
T_3 = Thickness of layer 3 expressed as a percentage of the total well thickness.
T_4 = Thickness of layer 4 expressed as a percentage of the total well thickness.

2.2.2. Geophysics

Geophysical prospecting has been used in previous works as a ground thermal conductivity estimator. The principal basis of these studies is the correlation of a geophysical parameter and thermal conductivity measurements (using KD2 Pro device) to finally predict the thermal behavior of the ground in depth. A more detailed description of these methods and their implementation in the area of the present research is included in the following subsections.

(1) Seismic data:

The first geophysical method makes reference to the implementation of seismic prospecting tests. In a previous research work, the mentioned tests were implemented on three different geological formations (schists, medium grain and coarse-grained adamellites) using MASW and seismic refraction techniques in order to register P and S waves velocities. At the same time, thermal conductivity of each formation was measured by the use of KD2 Pro equipment. These tests were made on the most and least decomposed samples of each geological environment to find the lowest and highest conductivity values. Finally, this published research correlates the propagation velocities of P and S waves and the thermal conductivity of samples from the same material [22].

The ultimate result of this work is to predict the thermal behavior of the geological formations included in the study. By identifying the propagation velocities of the seismic waves in a certain area, the evolution of the thermal conductivity of the ground in that area can be evaluated. Thus, 2D thermal conductivity sections provided in the mentioned research allow estimation of the evolution of ground thermal conductivity in depth for each specific formation.

In order to ensure application of this methodology, seismic refraction tests were conducted on the area where the borehole of study is located. The results of these tests are provided in Figure 3.

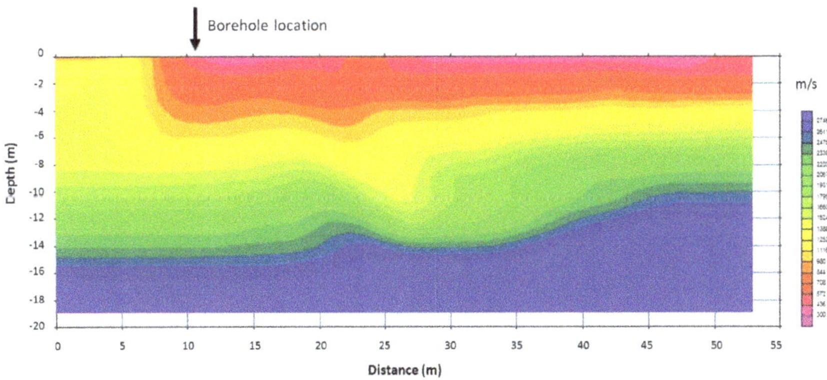

Figure 3. P-wave velocity distribution in the study area from seismic refraction tests.

After the distribution of the P wave velocity was identified in depth, different thermal conductivity measurements were taken in order to identify the most and least thermal conductive samples, meaning those with the highest and lowest compaction levels. These values correspond to the minimum thermal conductivity of anthropogenic fills and the maximum thermal conductivity for the altered adamellite;

they are presented in Table 3. It should be noted that altered adamellites were extracted from the drilling process at depths where they were identified. These samples were then used in thermal conductivity characterization.

Table 3. Highest and lowest thermal conductivity values detected for the formations constituting the borehole under study.

	Geological Formation	Thermal Conductivity * (W/mK)
Minimum value	Anthropogenic fills	1.105
Maximum value	Altered adamellite	2.672

* Thermal conductivity measuring was made by the use of KD2 Pro equipment.

By measuring P wave velocity and thermal conductivities in the study area, the correlation between both parameters was obtained (graphically presented in Figure 4). This is based on pairing the lowest thermal conductivity value with the lowest p wave velocity (in the same area) and the highest thermal conductivity with the highest p wave velocity. More information on this method is provided in the mentioned published research.

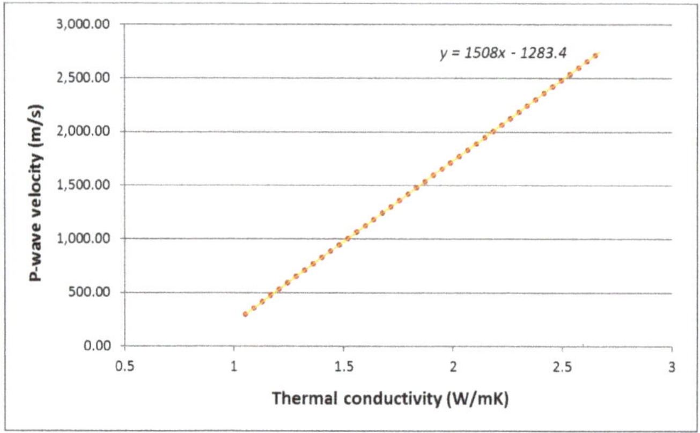

Figure 4. Thermal conductivity versus P-wave velocity in the study area.

From the above correlation and by following the instructions of the mentioned research, the distribution of the thermal conductivity parameter in the area considered here is displayed in the 2D section of Figure 5.

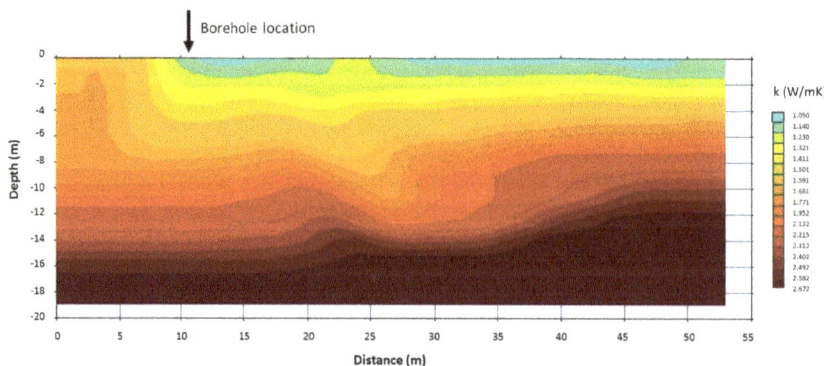

Figure 5. 2D thermal conductivity section in the area where the borehole of this research is located.

According to Figure 5, the volume of ground included under the borehole is constituted by a set of layers with different thermal conductivity values. This information can be observed in Table 4.

Table 4. Different thermal conductive layers identified in the volume of ground located under the borehole identified for this research.

	Thickness (m)	Thermal Conductivity (W/mK)
Layer 1	1.2	1.140
Layer 2	1.1	1.230
Layer 3	1.3	1.321
Layer 4	1.35	1.411
Layer 5	2	1.501
Layer 6	0.8	1.591
Layer 7	0.9	1.681
Layer 8	0.9	1.771
Layer 9	1	1.952
Layer 10	0.9	2.132
Layer 11	1.4	2.215
Layer 12	0.9	2.312
Layer 13	0.7	2.402
Layer 14	0.8	2.492
Layer 15	0.9	2.582
Layer 16 *	26.85	2.672

* From layer 16, the same thermal conductivity value is assumed until the end point of the drilling (43 m).

Finally, the global thermal conductivity of the borehole column is deduced from the application of the above data (Table 4) in Equation (1).

(2) Electrical resistivity:

In this case, electrical resistivity data were collected to finally create a 3D thermal conductivity map of the area of interest. The fundamentals of this method are similar to the one explained before; electrical resistivity results are correlated with thermal conductivity measurements and a relation between both parameters is obtained for a certain geological formation. The research work, including this methodology, was focused on granite rocks (adamellites), and the electrical resistivity was obtained using the Electrical Resistivity Tomography (ERT) technique. Thermal conductivity measurements were, in turn, taken using KD2 Pro equipment following the same operational procedure (tests were made on the most and least decomposed rocky samples) [23].

The results of this research disclose a certain relation between thermal conductivity and electrical resistivity. This relation can be observed in Equation (2).

$$k - 2{\cdot}10^{-7}x^2 + 0.0001x + 1.4881 \tag{2}$$

where:

k = thermal conductivity (W/mK)
x = electrical resistivity (Ω·m)

To apply Equation (2), the electrical resistivity of the materials in the study area must be known. To this end, an ERT test was conducted around the mentioned area, obtaining a 2D electrical resistivity section (presented in Figure 6).

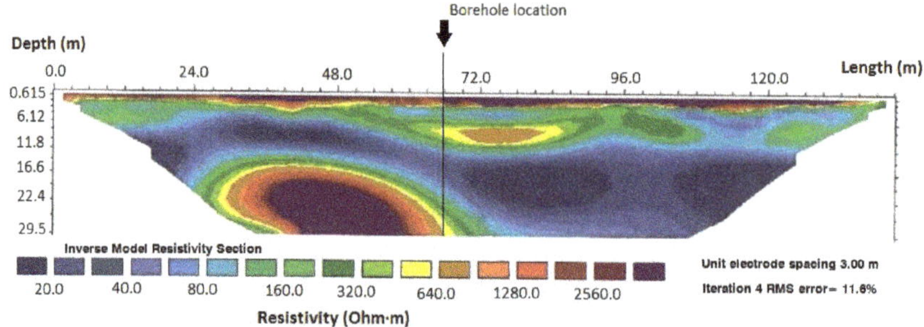

Figure 6. 2D electrical resistivity tomography in the area under study.

On interpretation of the above 2D electrical resistivity section, the borehole considered in this study is constituted by a series of layers with different thickness and characterized by variable electrical resistivity values. All these data are included in Table 5; the thermal conductivity of each layer is obtained by application of Equation (2).

Table 5. Layers detected in the borehole under study, according to the interpretation of ERT results.

	Thickness (m)	Electrical Resistivity (Ohm·m)	Thermal Conductivity (W/mK)
Layer 1	1	1280	1.943
Layer 2	4.12	100	1.498
Layer 3	6.88	450	1.574
Layer 4	10	55	1.494
Layer 5	7.5	360	1.550
Layer 6 *	13.5	2500	2.988

* Layer 6 is estimated based on the well logging of Figure 2.

As in the previous cases, Equation (1). must be used to finally define the global thermal conductivity of the borehole column from partial thermal conductivity values and thickness of each layer.

2.2.3. Thermal Response Test

The last procedure implemented in this research is the realization of a Thermal Response Test in the borehole. These tests are routinely used to estimate borehole thermal properties with regard to the mentioned thermal conductivity. The conventional TRT consists of circulating heated fluid (usually water) in a closed loop. During the test, fluid temperatures are measured at the ground heat exchanger inlet and outlet, along with the flow rate. Theses measured values are then analyzed by analytical or numerical models with the aim of calculating thermal conductivity and borehole thermal resistance [3,24].

(1) Test implementation

First, the borehole was geothermally prepared for the test by installing a polyethylene single-U tube heat exchanger of 32 mm with spacers located one meter apart. Taking advantage of the high groundwater level in the area, grouting material was not used [25,26]. The working fluid was water (during the test, low ambient temperatures were not expected) and the connection of the inlet and outlet heat exchangers and the TRT device was made with polyethylene tubes that were externally insulated. In order to set the initial condition of this test, a temperature register (PCE-T recorder) was used to measure the base temperature of the ground, obtaining a constant value of 14.6 °C at a depth of 40 m.

In this research, TRT was done according to UNE-EN ISO 17628:2017 regulations [27]. The TRT device implemented here constituted of a heat injection system, a circulating pump, and electrical resistance as heat source. The resistance allows three different heating levels, corresponding to the injection of 3 kW (stage 1), 6 kW (stage 2), and 9 kW (stage 3). The TRT equipment also included a Kamstrup energy meter (to register a large number of parameters), commercially known as MULTICAL 801.

Once the borehole was properly equipped, the sequence of events was as follows:

- Circuit filling and establishment of the appropriate working pressure.
- Activation of the circulating TRT pump and starting of the first heating stage (3 kW).
- General system operation during a certain period of time.
- Downloading and data management from the Kamstrup register.
- Calculation of the global thermal conductivity parameter.

The TRT duration is a controversial subject—while reducing TRT duration could help reduce costs, the accuracy of results could be affected. Following the regulation mentioned before [27], the minimum duration of the TRT can be estimated by Equation (3).

$$t\ (s) = \frac{5{\cdot}r^2}{\alpha} \tag{3}$$

where:

r = borehole radius (m)

$$\alpha = \frac{k_e}{c_v}$$

k_e = estimated thermal conductivity (W/mK)
c_v = volumetric thermal capacity (J/m^3/K)

By applying Equation (3) and estimating thermal conductivity of 1.80 W/mK and volumetric thermal capacity of 2.16 × 10^6 J/m^3/K [28], the minimum duration required for the thermal response test in the studied borehole would be:

$$t\ (s) = \frac{5{\cdot}0.11^2}{8.3{\cdot}10^{-7}} = 72891.\,56\ s\ \rightarrow\ 20.25\ \text{h}$$

Despite this value, the real duration of the test was 43 h, which sought to guarantee total stabilization of the system. Additionally, Figure 7 shows the TRT device and some sequences of the test.

(2) Thermal conductivity calculation

In a borehole heat exchanger of sufficient length in comparison with its radius, the analytical solution of Kelvin's Line Source can be applied to solve the heat equation and analyze TRT data. According to the infinite line-source model (use as a laboratory method since 1905), the thermal conductivity parameter can be obtained from the constant power rate and the slope of the temperature variation in time [29,30]. The interpretation of TRT results relies on a first-order approximation to linearize the mentioned infinite line-source model, neglecting the early measurements.

$$k = \frac{Q}{b{\cdot}4{\cdot}\pi{\cdot}H} \tag{4}$$

where:

Q = heat flux (kW/min)
b = slope (min)
H = borehole length (m)

Figure 7. Thermal response test in the studied borehole. Left: TRT device and Kamstrup register; right: TRT connected to the borehole heat exchanger.

3. Results and Discussion

3.1. Previous Methods Results

Thermal conductivity results of each method considered in this research are shown in Table 6. These results are obtained by the application of the stages described for each individual procedure. The methodologies belong to validated and already published researches. Consequently, the validity of the mentioned results is guaranteed.

Table 6. Thermal conductivity results of each method considered in this research.

Methodology	Thermal Conductivity (W/mK)
KD2 Pro	1.955
Seismic prospecting	2.337
Electrical resistivity	1.997

3.2. TRT Results

In addition to the thermal conductivity results deduced from the alternative methodologies, the TRT also provided a thermal conductivity value that will be compared with the ones in Table 6. After the corresponding operation of the TRT during the established period of time (43 h), inlet (T1) and outlet (T2) temperatures were registered (shown in Figure 8). It should be mentioned that the low temperature difference between T1 and T2 (displayed in Figure 8) is derived from the fact that the borehole length is only 43 m.

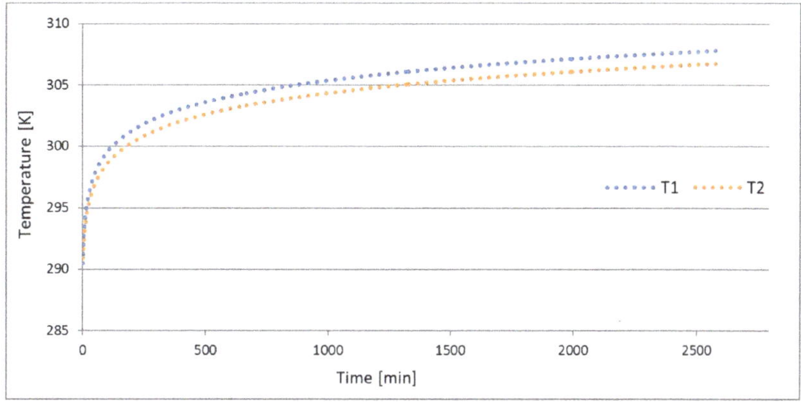

Figure 8. Evolution of inlet and outlet temperatures registered during the TRT.

From these results, the linear approximation required for the calculation of the thermal conductivity parameter was made for the period of time up to 1000 min (discarding the early measurements) in each temperature register. Figure 9 presents the equation of each linear approximation, consequently using the slope of these lines in corresponding thermal conductivity calculations. As shown in Figure 9, the interpretation of the TRT and the subsequent calculation of the thermal conductivity parameter is made by measuring the inlet and outlet fluid temperatures for time up to 1000 min.

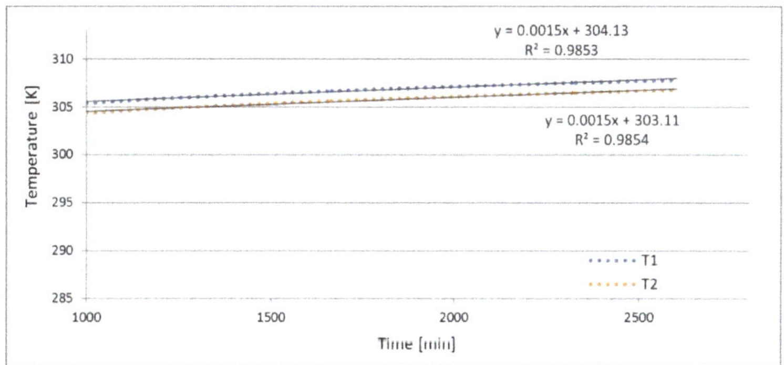

Figure 9. Equation of the linear approximation to temperature registers from time = 1000 min.

On observing Figure 9, we note that the slope of the linear approximation is the same for T1 and T2, such that the calculation of the thermal conductivity parameter is identical for both cases. When applying Equation (4), the following values were considered: $b = 0.0015$ min, $H = 43$ m and $Q = 1.875$ kW/min (resistance first stage (3000 kW)/time of the linear approximation (1600 min)). Thus, the global thermal conductivity of the borehole from TRT results takes a value of 2.313 W/mK.

3.3. General Comparison

Figure 10 graphically displays the results of each methodology considered in this research. It shows strong agreement in results between the different methods.

It is thus convenient to include the accuracy error of each of the methodologies shown in Figure 10: 10% for KD2 Pro, 14.2% for seismic prospecting, 16.7% for electrical resistivity, and 5% for TRT [4,31–33].

Considering TRT thermal conductivity value as the most accurate one and taking into account Figure 10, the following statements can be made:

- Thermal conductivities obtained by the alternative techniques are in strong agreement with the TRT result. The seismic prospecting method provides the most similar value, with a difference of only 0.024 W/mK with respect to the TRT value.
- The use of electrical resistivity tomography also allows to obtain thermal conductivity values close to the TRT result. In this case, the difference between both methods is 0.316 W/mK.
- The least accurate method is the use of the thermal conductivity map obtained by in situ KD2 Pro measurements. Despite having the least accuracy of all the procedures considered here, the difference with respect to the TRT is 0.358 W/mK.
- By evaluating the mentioned differences in terms of percentage, the errors of each alternative methodology in comparison with the TRT are 15.48% for the thermal conductivity map, 1.04% for seismic prospecting, and 13.66% when applying electrical resistivity tomography.

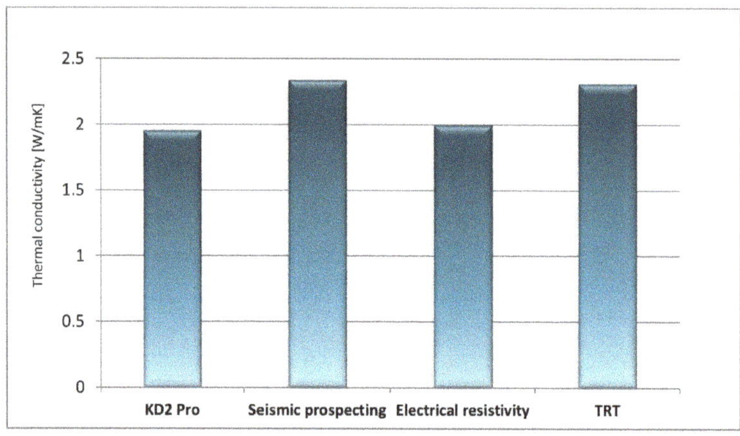

Figure 10. Thermal conductivity results of the implementation of the methodology used in this study.

4. Conclusions

TRT has been traditionally considered an appropriate technique to provide accurate thermal conductivity values of a borehole column. However, the high costs of these tests usually prompt researchers to look for alternative solutions that help characterize ground thermal behavior. This research integrates different existing methodologies to evaluate their validity through the results of a TRT made on the same study area. Provided that TRT is not always viable and based on the final results of this work, the most recommendable technique to be applied is seismic prospecting. It has been proved that this procedure is capable of providing highly accurate thermal conductivity values with errors below 1.5%. The high ground water level of the area may be a cause of deviation from the electrical and KD2 Pro methods, due to their sensitivity to this factor.

The remaining methodologies evaluated in this research could also be appropriate solutions to obtain approximate ground thermal conductivity. In the absence of a TRT or seismic profiles, the thermal conductivity map or electrical resistivity tomography could be of great help in ground thermal conductivity characterization. Despite the obvious advantages of TRT, the deep local nature of this test could be mitigated by using geophysical methods, as the ones presented in this study. The estimation of this parameter will be incredibly useful for the corresponding geothermal design, adjusting the number of boreholes and the total drilling length required in the shallow geothermal system. In view of the importance of identifying ground thermal conductivity in a GSHP system, the conclusions of this work are highly significant in the geothermal field.

Author Contributions: All authors performed the experimental and theoretical basis of the research. C.S.B., A.F.M. and I.M.N. implemented the methodology and calculations and analyzed the results. D.G.-A. provided technical and theoretical support. P.C.G. carried out the geophysics tests. C.S.B. wrote the manuscript and all authors read and approved the final version.

Funding: This research received no external funding.

Acknowledgments: The authors would like to thank the Department of Cartographic and Land Engineering of the Higher Polytechnic School of Avila, University of Salamanca, for allowing us to use their facilities and their collaboration during the experimental phase of this research. The authors also want to thank the Ministry of Education, Culture and Sport for providing a FPU Grant (Training of University Teachers Grant) to the corresponding author of this paper, making possible the realization of this work.

Conflicts of Interest: The authors declare no conflict of interest.

References

1. Ally, M.R.; Munk, J.D.; Baxter, V.D.; Gehl, A.C. Exergy analysis of a two-stage ground source heat pump with a vertical bore for residential space conditioning under simulated occupancy. *Appl. Energy* **2015**, *155*, 502–514. [CrossRef]
2. Han, C.; Yu, X.B. Performance of a residential ground source heat pump system in sedimentary rock formation. *Appl. Energy* **2016**, *164*, 89–98. [CrossRef]
3. Pasquier, P. Interpretation of the first hours of a thermal response test using the time derivative of the temperature. *Appl. Energy* **2018**, *213*, 56–75. [CrossRef]
4. Henk, J.L. Witte, Error analysis of thermal response tests. *Appl. Energy* **2013**, *109*, 302–311.
5. Pasquier, P.; Zarrella, A.; Marcotte, D. A multi-objetive optimization strategy to reduce correlation and uncertainty for thermal response test analysis. *Geothermics* **2019**, *79*, 176–187. [CrossRef]
6. Bandos, T.V.; Montero, Á.; Fernández de Córdoba, P.; Urchueguía, J.F. Improving parameter estimates obtained from termal response tests: Effect of ambient air temperature variations. *Geothermics* **2011**, *40*, 136–143. [CrossRef]
7. Choi, W.; Ooka, R. Interpretation of disturbed data in thermal response tests using the infinite line source model and numerical parameter estimation method. *Appl. Energy* **2015**, *148*, 476–488. [CrossRef]
8. ASHRAE. Geothermal energy. In *ASHRAE Handbook Heating, Ventilating, and Air-Conditioning Applications*; American Society of Heating, Refrigerating and Air-Conditioning Engineers: Atlanta, GA, USA, 2007; pp. 32.1–32.30.
9. Gehlin, S.; Hellström, G. Comparison of four models for thermal response test evaluation. *ASHRAE Trans.* **2003**, *109*, 131–142.
10. Spitler, J.D.; Gehlin, S.E. Thermal response testing for ground source heat pump systems: An historical review. *Renew. Sustain. Energy Rev.* **2015**, *50*, 1125–1137. [CrossRef]
11. Beier, R.A.; Smith, M.D. Minimum duration of in-situ tests on vertical boreholes. *ASHRAE Trans.* **2003**, *109*, 475–486.
12. Bozzoli, F.; Pagliarini, G.; Rainieri, S.; Schiavi, L. Short-time thermal response test based on a 3-D numerical model. *J. Phys. Conf. Ser.* **2012**, *395*, 012–056. [CrossRef]
13. Bujok, P.; Grycz, D.; Klempa, M.; Kunz, A.; Porzer, M.; Pytlik, A.; Rozehnal, Z.; Vojčinák, P. Assessment of the influence of shortening the duration of TRT (thermal response test) on the precision of measured values. *Energy* **2014**, *64*, 120–129. [CrossRef]
14. Poulsen, S.; Alberdi-Pagola, M. Interpretation of ongoing thermal response tests of vertical (BHE) borehole heat exchangers with predictive uncertainty based stopping criterion. *Energy* **2015**, *88*, 157–167. [CrossRef]
15. Choi, W.; Kikumoto, H.; Choudhary, R.; Ooka, R. Bayesian inference for thermal response test parameter estimation and uncertainty assessment. *Appl. Energy* **2018**, *209*, 306–321. [CrossRef]
16. Witte, H.J.L.; van Gelder, G.J.; Spitler, J.D. In Situ Measurement of Ground Thermal Conductivity: The Dutch Perspective. *ASHRAE Trans.* **2002**, *108*, 859–867.
17. Freifeld, B.M.; Finsterle, S.; Onstott, T.C.; Toole, P.; Pratt, L.M. Ground surface temperature reconstructions: Using in situ estimates for thermal conductivity acquired with a fiber optic distributed thermal perturbation sensor. *Geophys. Res. Lett.* **2008**, *35*, L14309. [CrossRef]
18. Sharqawy, M.H.; Said, S.A.; Mokheimer, E.M.; Habib, M.A.; Badr, H.M.; Al-Shayea, N.A. First in situ determination of the ground thermal conductivity for borehole heat exchanger applications in Saudi Arabia. *Renew. Energy* **2009**, *34*, 2218–2223. [CrossRef]

19. Geological and Mining Institute of Spain (IGME). *Geological National Mapping (MAGNA)*; IGME: Madrid, Spain, 2018.
20. Chamorro, C.R.; García-Cuesta, J.L.; Mondéjar, M.E.; Linares, M.M. An estimation of the enhanced geothermal systems potential for the Iberian Peninsula. *Renew. Energy* **2014**, *66*, 1–14. [CrossRef]
21. Blázquez, C.S.; Martín, A.F.; Nieto, I.M.; García, P.C.; Pérez, L.S.S.; Aguilera, D.G. Thermal conductivity map of the Avila region (Spain) based onthermal conductivity measurements of different rock and soil samples. *Geothermics* **2017**, *65*, 60–71. [CrossRef]
22. Blázquez, C.S.; Martín, A.F.; García, P.C.; González-Aguilera, D. Thermal conductivity characterization of three geological formations by the implementation of geophysical methods. *Geothermics* **2018**, *72*, 101–111. [CrossRef]
23. Nieto, I.M.; Martín, A.F.; Blázquez, C.S.; Aguilera, D.G.; García, P.C.; Vasco, E.F.; García, J.C. Use of 3D electrical resistivity tomography to improve the design of low enthalpy geothermal systems. *Geothermics* **2019**, *79*, 1–13. [CrossRef]
24. Raymond, J.; Therrien, R.; Gosselin, L. Borehole temperature evolution during thermal response tests. *Geothermics* **2011**, *40*, 69–78. [CrossRef]
25. Andersson, O.; Gehlin, S. State of the Art: Sweden, Quality Management in Design, Construction and Operation of Borehole Systems. Available online: http://media.geoenergicentrum.se/2018/06/Andersson_Gehlin_2018_State-of-the-Art-report-Sweden-for-IEA-ECES-Annex-27.pdf (accessed on 20 April 2019).
26. Gustafsson, A.-M.; Westerlund, L.; Hellström, G. CFD-modelling of natural convection in a groundwater-filled borehole heat exchanger. *Appl. Therm. Eng.* **2010**, *30*, 683–691. [CrossRef]
27. UNE-EN ISO 172628:2017. *Geotechnical Investigation and Testing. Geothermal Testing. Determination of Thermal Conductivity of Soil and Rock Using a Borehole Heat Exchanger*; International Organization for Standardizatio: Geneva, Switzerland, 2017.
28. Bates, D.R.; Estermann, I. *Advances in Atomic and Molecular Physics*; Elsevier: Amsterdam, The Netherlands, 1996.
29. Ingersoll, L.R.; Plass, H.J. Theory of the ground pipe heat source for the heat pump. *Heat. Pip. Air Cond.* **1948**, 119–122.
30. Carslaw, H.S.; Jaeger, J.C. *Conduction of Heat in Solids*, 2nd ed.; Clarendon Press: Oxford, UK, 1959.
31. *Decagon Devices*; KD2 Pro; Decagon Devices: Pullman, WA, USA, 2016.
32. *W-GeoSoft*; Geophysical Software; W-GeoSoft: Bioley-Orjulaz, Switzerland, 1988.
33. *Geotomo Software*; Geotomo Software: Penang, Malaysia, 2019.

Article

Serial Laboratory Effective Thermal Conductivity Measurements of Cohesive and Non-cohesive Soils for the Purpose of Shallow Geothermal Potential Mapping and Databases—Methodology and Testing Procedure Recommendations

Aleksandra Łukawska *, Grzegorz Ryżyński and Mateusz Żeruń *

Polish Geological Institute–National Research Institute, Rakowiecka 4 Street, 00-975 Warsaw, Poland; grzegorz.ryzynski@pgi.gov.pl

* Correspondence: aleksandra.lukawska@pgi.gov.pl (A.Ł.); mateusz.zerun@pgi.gov.pl (M.Ż.)

Received: 29 December 2019; Accepted: 15 February 2020; Published: 18 February 2020

Abstract: The article presents the methodology of conducting serial laboratory measurements of thermal conductivity of recompacted samples of cohesive and non-cohesive soils. The presented research procedure has been developed for the purpose of supplementing the Engineering–Geology Database and its part–Physical and Mechanical Properties of Soils and Rocks (abbr. BDGI-WFM) with a new component regarding thermal properties of soils. The data contained in BDGI-WFM are the basis for the development of maps and plans for the assessment of geothermal potential and support for the sustainable development of low enthalpy geothermal energy. Effective thermal conductivity of soils was studied at various levels of water saturation and various degrees of compaction. Cohesive soils were tested in initial moisture content and after drying to a constant mass. Non-cohesive soils were tested in initial moisture, fully saturated with water and after drying to a constant mass. Effective thermal conductivity of non-cohesive soils was determined on samples mechanically compacted to the literature values of bulk density. Basic physical parameters were determined for each of the samples. In total, 120 measurements of thermal conductivity were carried out, for the purposes of developing the guidelines which allowed statistical analysis of the results. The results were cross-checked with different measuring equipment and with the literature data.

Keywords: thermal properties of soil; transient line source method; ground source heat exchangers; geological mapping and databases; low enthalpy geothermal energy

1. Introduction

Thermal conductivity coefficient values of soil are useful in many subjects connected with energetics. Dynamic development of renewable energy sources and increasing awareness of necessity to reduce the use of fossil fuels are main trends that drive rising interest on thermal parameters of soil and rocks. Recognition of soil's thermal properties and parameters is essential when it comes to proper designing and building installations that use geothermal heat for energetics purposes, e.g., ground source heat pumps, [1–3]. At the same time, incorporating into the buildings thermally active construction elements that use thermal potential of a soil–rock medium and exchange heat between the ground and building is getting more and more popular. Few examples of those constructions are thermoactive foundations, foundation piles or tunnel walls [4–6].

Thermal properties of soil play key role also in determining possibilities of heat transfer in case of underground transmission infrastructure like high-voltage cables or district heating systems [7] as well as in terms of radioactive waste storage [8].

In this article authors present a developed version of quick and efficient methodology that allows conducting serial measurements of thermal properties of cohesive and non-cohesive soil. The methodology was formulated in Polish Geological Institute–National Research Institute (PGI-NRI) Soil and the Rock Testing Centre and presented in VI Polish Nationwide Geothermal Congress [9].

The Soil and Rock Testing Centre performs a number of tests of physical and mechanical parameters of different types of soil. Analyses are conducted in compliance with polish and international geotechnical standards [10–20]. Results of the tests performed for the purposes of tasks that fall into the scope of Polish Geological Survey activities, including the Engineering–Geological Database (BDGI), become a part of BDGI-WFM (physical and mechanical properties) component, that gathers information about soil and rocks tested in the laboratory.

Rising interest in ground heat pumps forces the need to recognize thermal parameters of soil and rocks. Current trends together with poor amount of information available in literature determine the need to develop a methodology that allows to make large amount of valid, repetitive tests in the shortest possible time. The presented methodology can find its use in Engineering–Geological and geotechnical laboratories. Thermal parameters presented in this paper were acquired with non-steady-state (transient) method with the use of thermal needle (hot wire method). During preparing the methodology, in order to verify the results, samples were also tested with transient plane source (hot disk) method, which also draw upon the non-steady-state (transient) method.

In 2014–2017, "Engineering–Geological database of physical and mechanical parameters of main soil and rocks lithotypes in Poland in terms of regional criterion (BDGI-WFM)" project was completed. The Engineering–Geological Database (BDGI) contains over 404,000 boreholes, including:

- A total of 308,120 boreholes located in areas of the Engineering–Geological Atlases (16 Polish urban agglomeration areas);
- A total of 19,860 boreholes all over the country (outside the Atlases area), introduced from the geological and geotechnical reports created since 2013;
- Over 76,020 boreholes from Warsaw in the form of Geotechnical Excavation Cards.

As a part of the project an Engineering–Geological Database that gathers data of physical and mechanical parameters of soil and rocks in Poland was established. BDGI-WFM contains information about location and contractor of the borehole/outcrop, timestamp and number of collected samples, their sampling category and measured parameters together with applied categorization standards. BDGI-WFM database is an integral part of Engineering–Geological Database (BDGI) and therefore can be updated on an ongoing basis using data from studies made for BDGI purposes. As part of the BDGI-WFM task, the following records were entered into the database:

- A total of 65,834 physical and mechanical parameters of soils;
- A total of 100 physical and mechanical parameters of rocks [21].

In 2018–2021, BDGI-WFM will be further extended by 120,000 parameters—the results of physical and mechanical laboratory tests—and enriched in the thermal properties of soil and rocks [22,23]. The thermal parameter data of soil and rocks gathered in databases like BDGI-WFM will be the basis for constructing geothermal 3D models and for low-temperature geothermal potential mapping [24].

2. Materials and Methods

2.1. Measurement Theoretical Background

Heat conduction (thermal conduction) is a physical phenomenon which involves the transfer of energy between adjacent molecules and electrons in the conducting medium. Energy is transported from places with a higher temperature to places with a lower temperature. In gases and liquids, energy is transferred mainly through inert collisions of molecules. In solids, transport occurs with the participation of free electrons and as a result of vibrations of the atoms of the crystal lattice of the medium [25,26]. Thermal conduction has been described by Fourier's law—the heat transfer in medium is proportional to the temperature gradient measured along the heat flow direction (1D). The mathematical equation takes the form:

$$q = -\lambda \frac{\partial T}{\partial x} \tag{1}$$

where:

q–local heat flux density ($W \cdot m^{-2}$),

λ–material's conductivity ($W \cdot m^{-1} \cdot K^{-1}$),

$\frac{\partial T}{\partial x}$–temperature gradient along the direction of heat flow ∇T ($K \cdot m^{-1}$).

Fourier's law applies in all states of matter, both for heat flow in steady state conditions (when the temperature difference(s) driving the conduction are constant) and heat flow in transient conditions (when the temperature changes in time during transfer period). In the first case, both the temperature and the heat flux density are functions of coordinates only (∂T, ∂x), in the second case both these values additionally depend on time.

The principle of thermal conductivity measurement in the steady state condition is to determine the heat flux Q migrating through the sample with known dimensions and to measure the temperature drop in the boundary of the sample with isothermal surfaces. The measurement under steady-state heat conduction is a relatively labor-intensive and time consuming process. The methodology uses complex control and regulation systems of the test stand. It is necessary to ensure perfect contact of the sample surface with the surfaces of the heater and cooler. In addition, proper thermal insulation of the other sample surfaces should be ensured. A constant value of the heat flow and a constant temperature of the sample-heater/cooler contact surface has to be preserved. Time period required to achieve steady state heat exchange in the sample (and thus to make measurements) is relatively long. The advantage of the thermal conductivity measurement under steady state conditions is the simplicity of the calculation formulas [27]. A steady-state heat flow is described by the equation:

$$\dot{q} = -\lambda F \frac{\partial T}{\partial x} \tag{2}$$

$\dot{q}$–heat flow (W),

F–cross-sectional area through which heat flows (m^2).

Measurement methods used to calculate thermal conductivity of a medium in a transient condition are based on the determination of the relationship between the density of the constant heat flux (emitted from a steady heat source) and the temperature changes over the time under non-steady energy flow through the tested material. Heat transfer under non-steady state conditions is described by the equation:

$$C_v \frac{\partial T}{\partial t} = \frac{\partial}{\partial x}\left(\lambda \frac{\partial T}{\partial x}\right) \tag{3}$$

C_v–volumetric heat capacity ($J \cdot m^{-3} \cdot K^{-1}$),

t–time (s).

For an isotropic medium, heat transport by conduction in one-dimensional space is described by the equation:

$$\frac{\partial T}{\partial t} = \alpha \frac{\partial^2 T}{\partial x^2} \tag{4}$$

where α ($m^2 \cdot s^{-1}$) is the ratio of the medium's thermal conductivity to its volumetric heat capacity (thermal diffusivity):

$$\alpha = \frac{\lambda}{C_v} \tag{5}$$

Measurement methods based on non-steady state are characterized by much faster test performance than steady state methods, are more versatile and easy to perform and characterized by relatively high measurement accuracy [27]. The presented methodology involves measurement of soil effective thermal conductivity in transient conditions by the thermal needle method.

Thermal needle method, also called the hot wire method, consists placing in tested sample a probe with a specific electrical resistance through which direct current of known intensity flows. Electric current flow increases temperature of the probe, which becomes a linear source of heat inside the sample. Heat flux generated during the test is constant. Measurement is carried out indirectly by determining heating rate of the tested sample [27]. As a result, it is possible to determine the increase or decrease of the temperature in the medium. Mathematical solution to the problem, based on the assumption of an infinitely long and infinitely thin heat source, was proposed by Carslaw and Jaeger [28]:

$$\Delta T = -\frac{Q}{4\pi\lambda}\, Ei\left(\frac{-r^2}{4\alpha t}\right) 0 < t \leq t_1 \tag{6}$$

$$\Delta T = -\frac{Q}{4\pi\lambda}\left[-Ei\left(\frac{-r^2}{4\alpha t}\right) + Ei\left(\frac{-r^2}{4\alpha(t-t_1)}\right)\right] t > t_1 \tag{7}$$

t–time from the beginning of heating (s),
ΔT–temperature rise since the beginning of heating (K),
Q–heat rate per unit of length ($W \cdot m^{-1}$),
r–radial distance from the line source (m),
α–thermal diffusivity ($m^2 \cdot s^{-1}$),
λ–thermal conductivity ($W \cdot m^{-1} \cdot K^{-1}$),
Ei–exponential integral,
t_1–heating time.
Exponential integral *Ei*, is defined by the series of expansion:

$$-Ei(-x) = \int_x^{\infty} \frac{e^{-x}}{x} \partial x \approx -\gamma - \ln x - \sum_x^{\infty} \frac{(-x)^n}{n \cdot n!} \tag{8}$$

where γ is Euler's constant (0.5772156649) $x = \frac{r^2}{4\alpha t}$

ASTM D5334-14 and IEEE Standards 442-2017 international thermal conductivity measurement standards suggest determination of thermal conductivity based on temperature data recorded for heating phase [19,20]. Interpretation method is based on the assumption that for a long measurement time t, when r is low and α is high, conditions outside $\ln x$ in the *Ei* series become negligibly small. Due to such assumptions, the equation can be presented as:

$$\Delta T \approx \frac{Q}{4\pi\lambda} \ln t + C \tag{9}$$

where C is a calibration constant. Therefore, when performing the interpretation proposed by ASTM D5334-14, the obtained results should be applied as a function of time on the logarithmic curve diagram. Finally, the thermal conductivity λ is calculated using the slope of the function:

$$\lambda = \frac{Q(\ln t_2 - \ln t_1)}{4\pi(\Delta T_2 - \Delta T_1)} \tag{10}$$

In fact, non-steady state thermal conductivity measurement never meets the assumptions made. Soil medium is not isotropic, the heat source is not infinitely long or thin, ambient temperature is never perfectly stable during the test—temperature drift always occurs. However, the authors have chosen this method because it is easier and much faster to perform than steady state methods, as well as because of its versatility and relatively high measurement accuracy as it was mentioned before.

2.2. Measuring Equipment

Portable, battery-operated KD2 Pro Thermal Properties Analyzer with TR-1 sensor was utilized to measure thermal conductivity of the samples. Large, single needle TR-1 sensor (2.4 mm diameter × 100 mm long) was chosen to perform the tests as the producer recommends it for soil and other granular materials [29]. Measuring range for thermal conductivity of TR-1 sensor lies between 0.10 and 4.00 $W \cdot m^{-1} \cdot K^{-1}$. Accuracy of the sensor is ± 10% from 0.2–4 $W \cdot m^{-1} \cdot K^{-1}$ and ± 0.02 $W \cdot m^{-1} \cdot K^{-1}$ from 0.1–0.2 $W \cdot m^{-1} \cdot K^{-1}$.

TK04 Thermal Conductivity Meter with disc-shaped probe for plane surfaces (Standard HLQ probe) was used to validate the results obtained with KD2 Pro Analyzer.

2.3. Auxiliary Equipment

- Stainless steel cylinder with a tapping fork that is used to determine non-cohesive soil bulk density was utilized to form sand samples for thermal conductivity analyses;
- Metal form with a rammer, that were utilized for molding cohesive soil samples;
- Knife with a straight edge was used to even samples surface to the level of upper edge of the mold;
- Laboratory thermometer to monitor environment temperature during measurements.

Measuring and auxiliary equipment is shown in Figure 1.

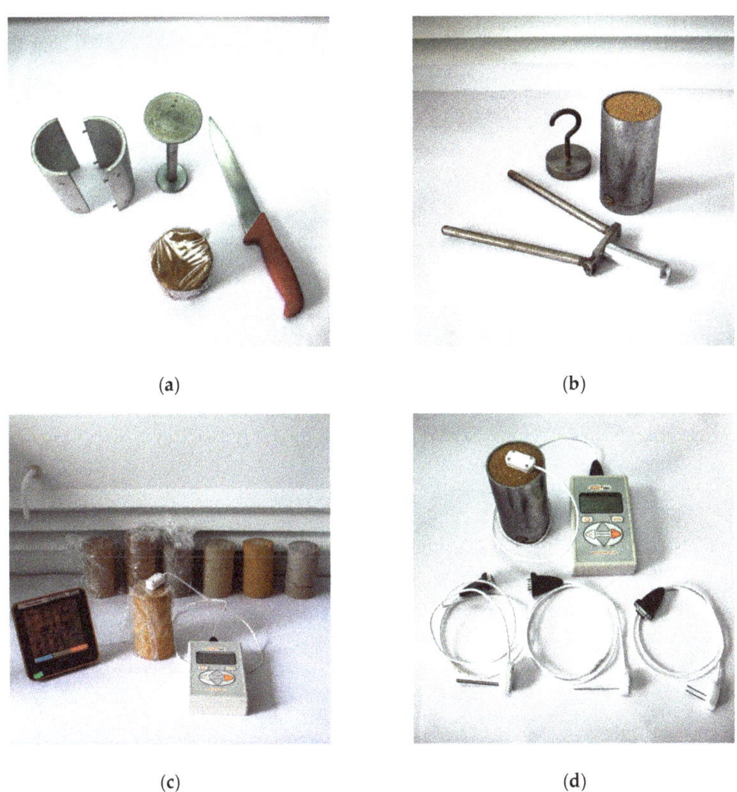

(a) (b) (c) (d)

Figure 1. Equipment for laboratory testing of thermal conductivity of soil: (**a**) Metal mold to form cylindrical samples of cohesive soil with a knife and rammer; (**b**) Steel cylinder with tapping fork that is used to form non-cohesive soil samples; (**c**) Molded cohesive soil samples during analysis; (**d**) A set of exchangeable thermal needle probes with KD2 Pro Thermal Properties Analyzer.

2.4. Samples Used in the Testing Procedure

For the purposes of collecting data for BDGI-WFM two sorts of samples are utilized: undisturbed (1st quality class of soil sample) and disturbed (3rd quality class of soil sample) according to EN 1997-2 Eurocode 7 [10].

Undisturbed samples are collected in a way that allows preserving their natural structure as well as full range of the soil's physical and mechanical features (A sampling category). Undisturbed samples truly represent in situ soil layer in terms of its properties. In case of undisturbed soil samples, measurement of thermal conductivity can be carried out without any additional preparation of the probe.

Disturbed samples (B category soil sampling) are most commonly used in laboratory testing procedures. Natural water content and particle size distribution remain unchanged, but the structure of the soil is damaged or modified during the sampling operation. Therefore, to obtain valid results of thermal conductivity, test molding of disturbed samples is required.

Depending on the sample quality class the process of sample preparation is carried out different way (Figure 2).

A set of three thermal conductivity measurements is performed on each sample. This kind of testing cycle allows the identification of invalid thermal conductivity values and enables carrying out statistical interpretation of the results.

Consecutive results should not differ more than 10%, otherwise they are not being taken into consideration.

In order to characterize physical properties of examined soil and to identify their possible influence on acquired results of thermal conductivity coefficient authors recommend performing undermentioned supplementary analyses:

- determination of water content,
- determination of particle size distribution,
- determination of bulk density for cohesive soils and relative density for non-cohesive soils,
- determination of liquid and plastic limits.

The authors recommend those tests to be performed according to international geotechnical standards, for example ISO 17892: Geotechnical investigation and testing—Laboratory testing of soil [13–15,17].

Please note that to conduct additional tests sampling properly, larger amounts of soil will be required.

2.5. Sample Preparation

Minimum mass required to perform the analysis according to proposed procedure should stand 1500 g and 1000 g minimum for cohesive and non-cohesive soil respectively. Cohesive soils should be examined in two conditions: wet (in their natural moisture content) and after drying to a constant mass. Non-cohesive soils should also be tested in three cases: initial water content, dried to a constant mass and fully saturated with water.

Because in case of cohesive soils thermal conductivity tests are performed on the samples with natural water content, all cohesive soil samples should always be hermetically packed (three layers of plastic zip-lock bags recommended) directly after sampling and stored in the refrigerator to prevent the loss of water content.

Container for sample preparation should be made of stiff and rust-resisting material (e.g., stainless steel). Container dimensions should make at least 5 to 12 cm.

Prepared samples should be placed in the examination room at least 24 h before starting the test so that the equilibrium between specimen and environment temperature is reached. It is recommended to keep the environment temperature as stable as it is possible. There should be no intensive air flows or humidity changes during the test as it is proven that this kind of fluctuations in environmental conditions can affect the results [29]. During the examination, wet samples should be protected from changes in moisture content by covering with stretch wrap. Temperature and humidity level should be registered during the test.

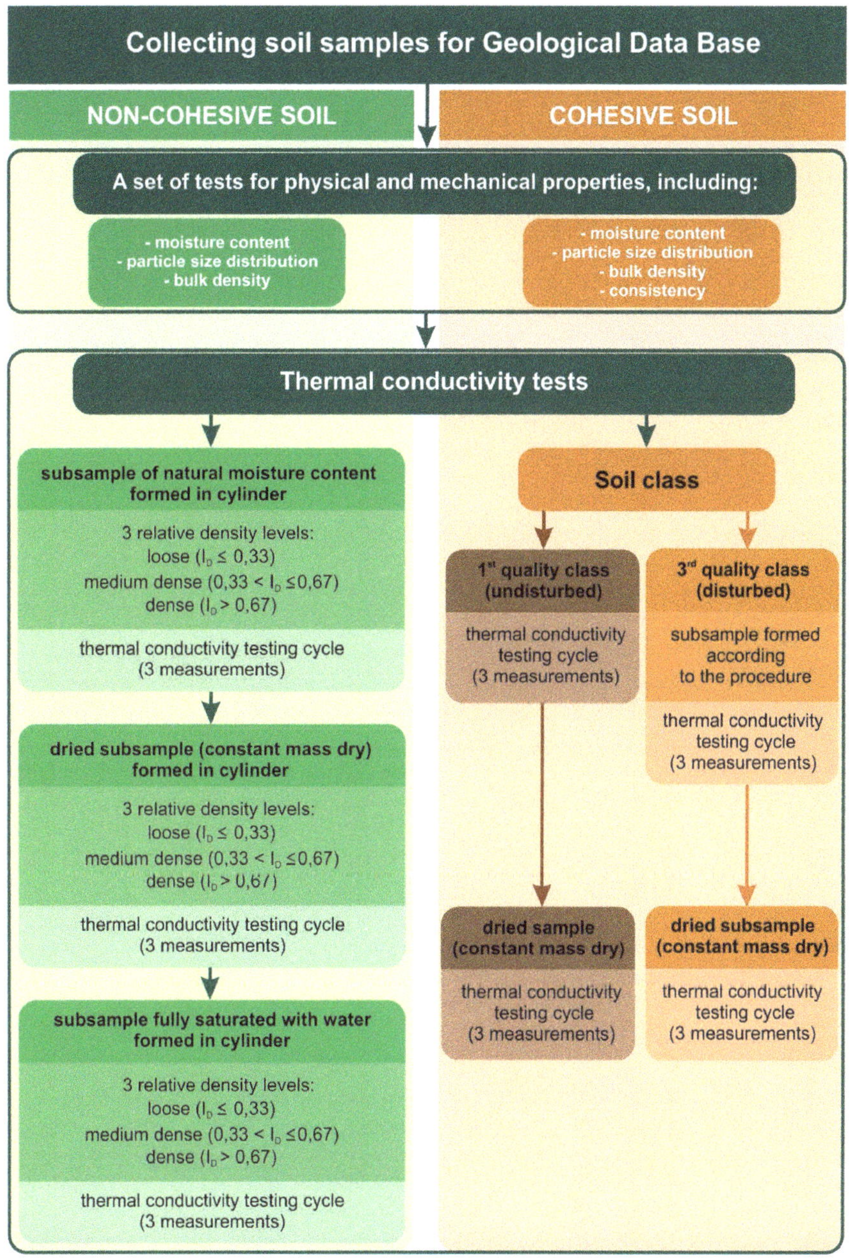

Figure 2. Scheme of effective thermal conductivity measurements in PGI-NRI. I_D stands for the density index.

2.6. Preparation of Cohesive Soil Samples

2.6.1. Samples with Natural Water Content

To prepare cohesive soils samples cylindrical mold made of stainless steel is being used (Figure 1a). Small lumps of soil are placed inside the form with layers and firmed to desired density with a metal rammer. The soil is then rammed to the point in which it will reach bulk density specific for the soil of the same water content in situ [30]. When the soil sample reaches the level of the upper ridge of the container, it is cut with a straight edge knife. Cutting is done from the center of the sample towards its rims. Molded sample is then removed from the form carefully, without intruding its previously molded structure. Please note, that the test can be also conducted directly in the sample container, which can be particularly useful in case of the soils of very soft consistency.

2.6.2. Dry Samples

Molded cohesive soil samples are placed in evaporating dish and dried to a constant mass in the laboratory drying oven for 72 h minimum. The drying oven used in this procedure should be capable of maintaining a temperature of 105 °C ± 5 °C [17]. When the sample reaches constant mass dry state, a hole is drilled longwise the sample, so that inserting thermal needle is possible. The drill should reach both sides of the sample [19]. The diameter of the drill should be as similar to the diameter of a needle probe used in the test (2.4 mm for TR-1 probe) as it is possible. Once drilled, the hole should be cleaned of the dust and filled with thermal grease to ensure good thermal contact between the sensor and tested soil.

2.7. Preparation of Non-cohesive Soil Samples

2.7.1. Samples with Natural Water Content

The principles of tapping fork test method that is used for determining the bulk density of non-cohesive soils was the base for developing the way of preparing non-cohesive soil samples for the purposes of this methodology. The soil's bulk density is determined by placing it in a metal cylinder of a specified dimensions and compacting the soil by putting it into vibrations with a metal tapping fork (Figure 1b). The compaction procedure is fully described in polish geotechnical investigation standard PN-B-04481:1988 [18].

According to the fore mentioned standard, non-cohesive soils are prepared in a stainless steel, flat-bottom cylinder. The cylinder is slowly filled with examined soil by using a laboratory funnel. The funnel should at first be leaned on the cylinder's bottom and as the cylinder was filled the funnel should be slowly elevated to let the soil form thin layers and therefore avoid closing air between soils particles. When the sample container is full, the specimen should be even with straight edge knife to the level of the upper rim of the cylinder. The cylinder is then topped with a steel ring on the surface of the soil and the soil is compacted to a desired bulk density by striking the sides of a cylinder with metal tapping fork. Water content is measured after finishing the compaction process.

2.7.2. Dry Samples

The sample should be dried to a constant mass in the laboratory drying oven for at least 12 h. The drying oven should be capable of maintaining a temperature of 105 °C ± 5 °C [17]. After drying, the specimen is gently grinded in a mortar to get rid of the aggregates that might have formed in the drying process. The authors recommend using rubber-headed pestle to grind the soil to avoid crushing the grains. Again, the soil is placed in the cylinder with a funnel. After filling the cylinder, the specimen's surface is cut with a straight edge knife to the level of the upper rim of the cylinder and compacted to desired bulk density as it is described in the Section 2.7.1.

2.7.3. Samples Fully Saturated with Water

To prepare non-cohesive soil samples fully saturated with water, first the water is poured to the cylinder to approximately 1/4 of its total volume. The soil is then evenly placed in the cylinder using a funnel. During the filling, water is gradually added in order to keep the specimen fully saturated. Whole sample volume should all the time be kept under the water level. The specimen is cut to the level of the upper rim of the cylinder and compacted to desired bulk density. Percentage water content is determined after the whole process.

2.8. Thermal Conductivity Test Procedure

Depending on the physical properties of tested material, different type of thermal probe should be chosen. For the purpose of testing soft soils, TR-1 sensor is recommended [29]. In case of coarse grained ($\phi > 2$ mm) non-cohesive samples using the TR-1 sensor is not possible due to a risk of damaging the sensor while embedding the probe into sample. For the purposes of compiling the described methodology, in order to measure coarse-grained samples (PNS-1, PNS-4) stiff large diameter (3.9 mm diameter and 60 mm long) the RK-1 needle probe was utilized instead.

Before starting each series of measurements, the sensor performance should be verified by measuring thermal conductivity of certified reference material provided by a producer. If the result of the verification is valid, the test procedure can be conducted on soil samples.

The needle sensor should be placed carefully in the center of the sample. A minimum of 2 cm of material should be allowed parallel to the sensor in all directions. The sensor should be inserted all the way into the measured soil. Once it is placed inside the specimen, the sensor position shouldn't be changed in order to provide best possible thermal contact between the sensor and measured medium. To allow samples and sensor to come to temperature equilibrium, it is advised to wait 15 min time before starting the measurement.

According to the methodology, three measurements were performed on each sample. Probe's location wasn't changed during or between the readings. Fifteen minutes pause applied between measurements was optimal for the sensor-sample temperature to equilibrate.

Verification of the Results

To verify validity of the results obtained in compliance with the above mentioned procedure the results were cross-checked using TK04 Thermal Conductivity Meter with disc-shaped 88-mm diameter standard HLQ probe presented in Figure 3a. The sensor's needle is embedded in the underside of the disc material. HLQ probe covers thermal conductivity range of 0.3 to 10 $W \cdot m^{-1} \cdot K^{-1}$ and can be used for measuring various plane surfaces [31]. Thermal conductivity is measured using transient hot wire method. Heat generated by the disk-shaped probe penetrates the sample and partially also the probe material (Figure 3b). TK04 software provided by a producer automatically corrects this effect to evaluate thermal conductivity of the sample.

Sand gravel sample was formed to a dense degree of compaction ($I_D > 0.67$) according to the afore-described procedure and thermal conductivity was measured. The same sample was then compacted to the same density using Proctor apparatus [11]. The results of both tests again were strongly consistent (PNS-1, Table 1). In both cases, thermal grease provided by the producer was utilized to provide a proper contact between the probe and examined specimen.

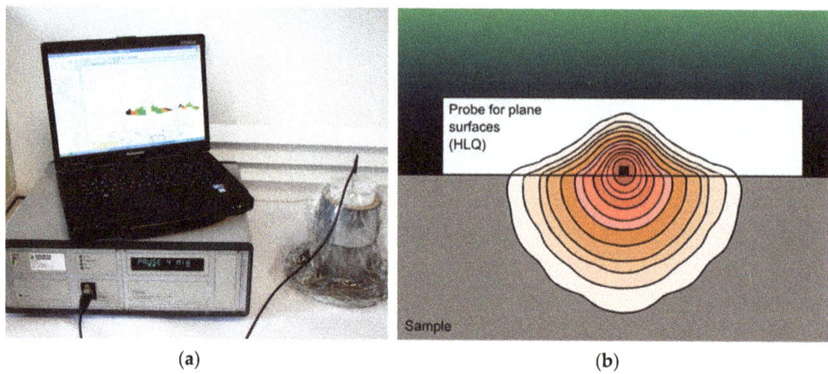

(a) (b)

Figure 3. (**a**) Non-cohesive soil sample during TK04 Thermal Conductivity Meter measurement with disc-shape probe; (**b**) heat generated during the test by a disc-shaped HLQ probe.

Table 1. Thermal conductivity and physical properties of tested non-cohesive soils.

Sample no	Soil Type	Particle Size Distribution [%]			Water Content	Relative Density	Bulk Density [g/cm^3]	Thermal Conductivity λ [W/(m*K)]	PORT PC Category
		>2 mm	>0.5 mm	>0.25 mm					
PNS-1[1]	Sand gravel	12.09	64.69	83.47	dry	dense	1.89	0.473 0.474 0.475	2
					natural water content	dense	1.97	2.814 2.817 2.834	3
					fully saturated	dense	2.22	3.227 3.270 3.293	4
					dry (validation)	dense	1.91	1.050 1.075 1.087	2
PNS-2	Medium sand	0.16	5.23	57.39	dry	loose	1.55	0.246 0.246 0.246	1
						medium dense	1.70	0.279 0.280 0.286	1
						dense	1.85	0.335 0.350 0.351	2
					natural water content	loose	1.79	2.392 2.406 2.407	3
						medium dense	1.83	2.523 2.541 2.542	3
						dense	1.87	2.630 2.634 2.640	3
					fully saturated	loose	1.84	2.822 2.826 2.827	4
						medium dense	2.05	2.920 3.013 3.038	4
						dense	2.16	3.112 3.122 3.147	4

Table 1. *Cont.*

Sample no	Soil Type	Particle Size Distribution [%]			Water Content	Relative Density	Bulk Density [g/cm³]	Thermal Conductivity λ [W/(m*K)]	PORT PC Category
		>2 mm	>0.5 mm	>0.25 mm					
PNS-3	Coarse sand	1.09	97.45	99.58	dry	dense	1.82	0.324 0.324 0.325	2
					natural water content	dense	1.87	1.871 1.886 1.888	3
					fully saturated	dense	1.98	2.732 2.739 2.715	4
PNS-4 [1]	Gravel	86.11	95.08	95.81	dry	dense	1.68	0.229 0.232 0.234	5
					natural water content	dense	1.73	1.878 1.826 1.827	5
					fully saturated	dense	2.12	2.831 2.866 2.878	6
PNS-5	Medium sand	0.03	7.33	55.91	dry	dense	1.71	0.364 0.363 0.363	2
					natural water content	dense	1.85	1.421 1.441 1.459	3
					fully saturated	dense	2.15	2.996 3.019 3.010	4
PNS-6	Fine sand	0.03	0.93	28.98	dry	dense	1.75	0.307 0.307 0.310	1
					natural water content	dense	1.81	1.096 1.096 1.109	3
					fully saturated	dense	1.97	2.875 2.875 2.927	4

[1] tested with TR-1 large diameter needle probe.

In case of cohesive soils KD2 Pro thermal conductivity verification test was performed on undisturbed silt sample with preserved natural structure [PS-1]. After the test the sample was pulled down, formed according to the procedure and thermal conductivity was measured again. Both tests showed very similar thermal conductivity coefficient values (see Table 2).

3. Results

The authors emphasize that the article contains only those results of thermal conductivity of soils that were obtained during the development of the methodology. The article is not intended to present the results of tests for individual soils.

Samples chosen for the present paper represent wide range of lithology types, classified on the basis of particle size distribution according to EN-ISO 14688 geotechnical standard [12]. Criterion of soil's regional origin was omitted in the procedure.

A set of samples collected for the BDGI-WFM purposes were used. Physical parameters of samples that are believed to influence thermal properties of the soil (water content, bulk density, particle size distribution, consistency index) were determined and taken into consideration.

- Water content was measured according to EN ISO 17892-1:2014, Geotechnical investigation and testing—Laboratory testing of soil—Part 1: Determination of water content [17].
- Bulk density was measured according to EN ISO 17892-2:2014, Geotechnical investigation and testing—Laboratory testing of soil—Part 2: Determination of bulk density [13].

- Particle size distribution was measured according to EN ISO 17892-4:2017 Geotechnical investigation and testing—Laboratory testing of soil—Part 4: Determination of particle size distribution [14].
- Determination of soil's consistency was based on methodology according to EN ISO 17892-12:2018 Geotechnical investigation and testing—Laboratory testing of soil—Part 12: Determination of liquid and plastic limits [15].
- Description and classification of soil was done according to ISO 14688-1:2017 Geotechnical investigation and testing—Identification and classification of soil—Part 1 and 2 [16].

On each sample three thermal conductivity measurements were performed. Overall 120 measurements of cohesive and non-cohesive soils were taken during preparation of this methodology.

All of thermal conductivity test results together with sample's physical parameters are given in Table 1 (non-cohesive soils) and Table 2 (cohesive soils).

Results obtained in the laboratory were compared with numerical data from the literature guidelines provided by Polska Organizacja Rozwoju Technologii Pomp Ciepła (Polish Organization of Heat Pump Technology Development) "PORT PC" [32]. PORT PC Guidelines include recommended thermal conductivity values for certain soil types for the design of ground source heat pumps (GSHP). Those values are presented in above-mentioned guidelines in tabular form as ranges of values with a certain value recommended for the GSHP design.

3.1. Non-cohesive Soils

Thermal conductivity tests were performed on six samples of non-cohesive soils: fine sand (PNS-6), coarse sand (PNS-3), gravel (PNS-4), gravel sand (PNS-1) and two samples of medium sand (PNS-5, PNS-2).

Each sample was measured in three different degrees of water saturation: fully saturated, constant mass dry and with initial water content. Samples were compacted to a "dense" state according to a relative dense term (density index $I_D \geq 0.67$) [14].

Overall thermal conductivity coefficient for 25 non-cohesive soils samples of different saturation degree and density index were determined.

In order to identify if there is any correlation of compaction rate and thermal conductivity of the soil one of the samples (PNS-2) was also thermally tested in three different degrees of compaction: loose ($I_D \leq 0.33$), medium dense ($0.33 \leq I_D \leq 0.67$) and dense ($I_D \geq 0.67$).

All of the thermal conductivity test results together with sample's physical parameters are given in Table 1.

Thermal conductivity results as a function of moisture content are shown in the Figure 4. It is clearly visible that thermal conductivity of the soil increases with increasing water content. The increase of thermal conductivity coefficient is more rapid in the small moisture content (0%–5%) and becomes slower in higher moisture contents. The highest thermal conductivity values were obtained in samples fully saturated with water (up to 3.3 $W \cdot m^{-1} \cdot K^{-1}$).

In the red ellipses the effect of medium sand compaction is visible. The compaction process causes decrease in moisture content, but increase in specimen's relative density gives higher values of thermal conductivity coefficient despite lower moistness.

Table 2. Thermal conductivity and physical properties of tested cohesive soils.

Sample no	Soil Type	Particle Size Distribution [%]				Water Content [%]	Subsample Type	Consistency	Bulk Density [g/cm^3]	Thermal Conductivity λ [W/(m*K)]	PORT PC Category
		>2 mm	2–0.5 mm	0.05–0.002 mm	<0.002 mm						
PS-1	Silt	0.00	29.89	60.33	9.78	<2	molded, dry	very stiff	1.79	0.809 0.828 0.833	7
						27.00	molded, natural water content	firm	2.11	2.139 2.141 2.143	8
							natural structure	firm	2.16	2.027 2.028 2.050	8
PS-2	Sandy clay	0.19	65.42	16.76	17.63	<2	molded, dry	very stiff	1.93	1.141 1.145 1.145	9
						18.34	molded, natural water content	stiff	2.15	2.547 2.564 2.586	10
PS-3	Sandy clay	0.46	58.48	21.28	19.78	<2	molded, dry	very stiff	2.01	1.421 1.423 1.475	9
						13.74	molded, natural water content	very stiff	2.18	2.566 2.571 2.578	10
PS-4	Sandy clay	0.21	56.35	25.24	18.2	<2	molded, dry	very stiff	1.97	1.266 1.270 1.283	9
						16.76	molded, natural water content	stiff	2.12	2.298 2.298 2.308	10
PS-5	Sandy clay	0.68	36.3	43.23	19.7	<2	molded, dry	very stiff	1.98	1.112 1.117 1.129	9
						16.37	molded, natural water content	stiff	2.11	2.065 2.107 2.111	10
PS-6	Clay	1.65	18.43	33.56	46.36	<2	molded, dry	very stiff	1.71	0.744 0.749 0.746	9
						22.86	molded, natural water content	stiff	2.1	1.861 1.875 1.880	10
						37.98	remolded, water added	soft	1.9	1.359 1.367 1.382	10
						51.21	remolded, water added	very soft	1.72	1.061 1.089 1.061	10

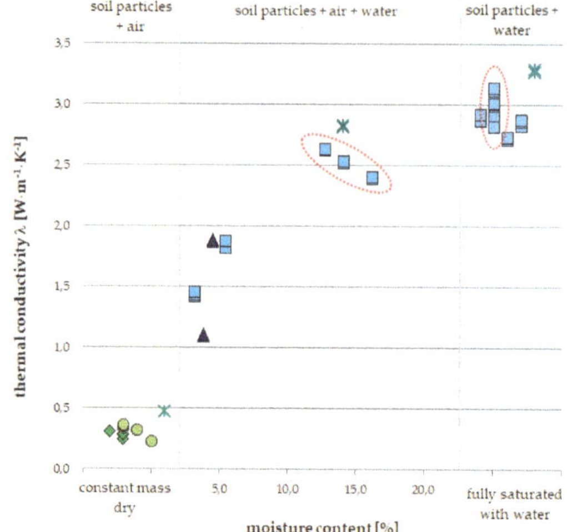

Figure 4. Thermal conductivity as a function of moisture content for non-cohesive soils. Red ellipses indicate the effect of compaction of single non-cohesive soil sample on thermal conductivity values.

3.2. Cohesive Soils

Thermal conductivity coefficient of 15 cohesive soils samples was measured. Each sample was examined in natural water content and in constant mass dry state. 14 samples were formed according to the procedure. Undisturbed silt sample (PS-1) was used to verify and evaluate the results using TK04 conductivity meter.

To determine the relationship between water content and thermal conductivity of the soil, one of the samples (PS-6) was tested fourfold: each time in different consistency, including constant mass dry. Plasticity of the clay was increased by mixing the sample with distilled water and stirring carefully until it formed homogeneous paste. The paste was then remolded to desired density and the testing cycle of three measurements was conducted.

Thermal conductivity results for cohesive soil samples presented in this paper as a function of moisture content are shown in Figure 5.

Similarly to non-cohesive soil, thermal conductivity of cohesive soil increases with increasing water content. The highest thermal conductivity values of cohesive soils are observed in plastic state. However, at some point of moistness adding more water to the sample has a reverse effect on measured thermal coefficient values (clay sample, inside dashed red ellipse). It is caused by the fact that the bulk density of a cohesive soil decreases at high water contents; the percentage content of water grows and the content of the soil's particles drops. As pores are getting filled with water, soil particles are losing contact with each other and heat transfer inside the sample becomes less efficient.

The highest thermal conductivity values were obtained for sandy clays (up to 2.6 $W{\cdot}m^{-1}{\cdot}K^{-1}$). The effect is believed to be caused by high quartz content in this type of soil. Clay samples present significantly lower thermal conductivity values (up to 1.9 $W{\cdot}m^{-1}{\cdot}K^{-1}$). It is a result of the lower thermal conductivity of clay minerals [26].

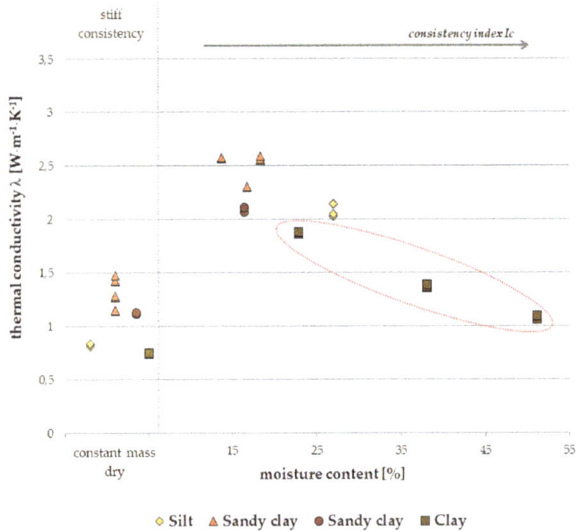

Figure 5. Obtained thermal conductivity values as a function of moisture content for cohesive soils. Red ellipse indicates the effect of increasing water content on thermal conductivity values of single cohesive soil sample.

Repeated thermal conductivity results at the same water content for clay and silt samples are consistent. For sandy clays (till) thermal conductivity values are less predictable. In constant mass dry state thermal conductivity coefficient for sandy clays ranges from 1.1 to 1.5 $W{\cdot}m^{-1}{\cdot}K^{-1}$.

It is believed that those variations are caused by a different proportions between >2mm, >0.5mm and >0.25mm soil particles fractions. Higher content of gravel, sand and silt fractions, which in Polish conditions are rich in quartz, is responsible for higher thermal conductivity values.

4. Discussion

Results obtained for the purposes of this paper were compared with data presented in "Guidelines for Ground Source Heat Pumps design. Part 1 Ground heat exchangers" published by Polish Organization for the Development of Heat Pump Technology (PORT PC), that is commonly used in Poland [32]. Thermal conductivity values presented in the guidelines come from VDI 4640 Blatt 2 2001 Thermal use of the underground–ground source heat pump systems and Earth Energy Designer software and do not cover all kind of soil found in Poland [33,34].

Ranges of thermal conductivity values provided by PORT PC Guidelines were presented in Figure 6 in graphical form, to give reference to results of thermal conductivity tests obtained in the article. Both groups of the results are mostly consistent; results of our measurements mostly fit in PORT PC recommended value ranges (Figure 6).

Some variations can be observed as well resulting from different way of soil classification than those provided in geotechnical standards. For example, categorization criteria available in literature put in a single category clays and sandy clays (till), while by analyzing the results of our laboratory measurements one can tell that thermal conductivity coefficient values for those two types of soil vary significantly. The authors think that more precise way of categorization the soil will be more useful. Classification of soil used in geotechnical practice identify the soils basing on the particle size grading (in case of coarse particles), plasticity of fine particles and the content of organic matter (for organic soils) [16]. It is crucial to take those properties into consideration, as each of those features can influence thermal properties of soils. Further classification criteria such as stratigraphy, regional geological conditions and genesis of the soil should be taken into account and will be the subjects of future researches.

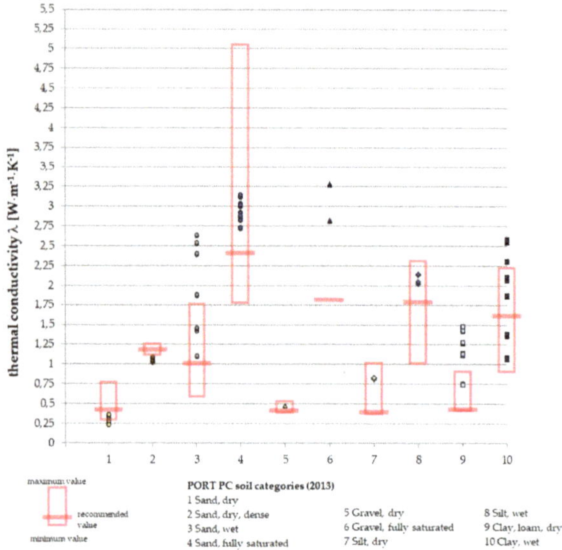

Figure 6. Obtained thermal conductivity results for different lithotypes in comparison with data from PORT PC guidelines.

Future research directions will comprise further filling the BDGI-WFM database with thermal conductivity parameters for different types of soils on a wide range of moisture contents. Collecting large amounts of data will allow constructing nomograms of thermal conductivity coefficient as a function of moisture content that will provide a direct tool to estimate thermal conductivity coefficient for the specific soil type (Figure 7).

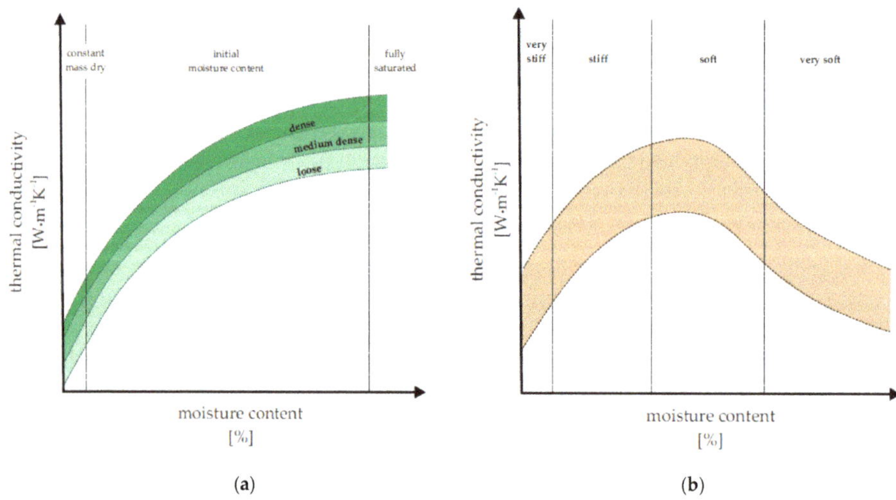

Figure 7. Conceptual model of nomograms of effective thermal conductivity coefficient as a function of moisture content for non-cohesive (**a**) and cohesive (**b**) soil.

As mentioned before, it is important for reliability of the data to always consult local geological conditions. Creating regional thermal properties databases that integrate data from the whole country/region should be one of the future concerns [35].

Figure 8 presents the shallow geothermal potential of Poland and the result of processing, analysis and interpretation of subsurface data from the thematic databases, atlases and serial maps gathered in the archives of the Polish Geological Institute National Research Institute. According to a computational algorithm the reclassification of lithological parameters into the geothermal parameters was performed as well as the values of geothermal conductivity λ [$W{\cdot}m^{-1}{\cdot}K^{-1}$] of analyzed rocks and soils were calculated. Based on the results of calculations spatial layer of average effective thermal conductivity coefficient λ was prepared for the depth interval up to 100 m. The results of performed statistical estimations allowed to create the synthetic characteristics of shallow geothermal energy potential for the major geological regions of Poland.

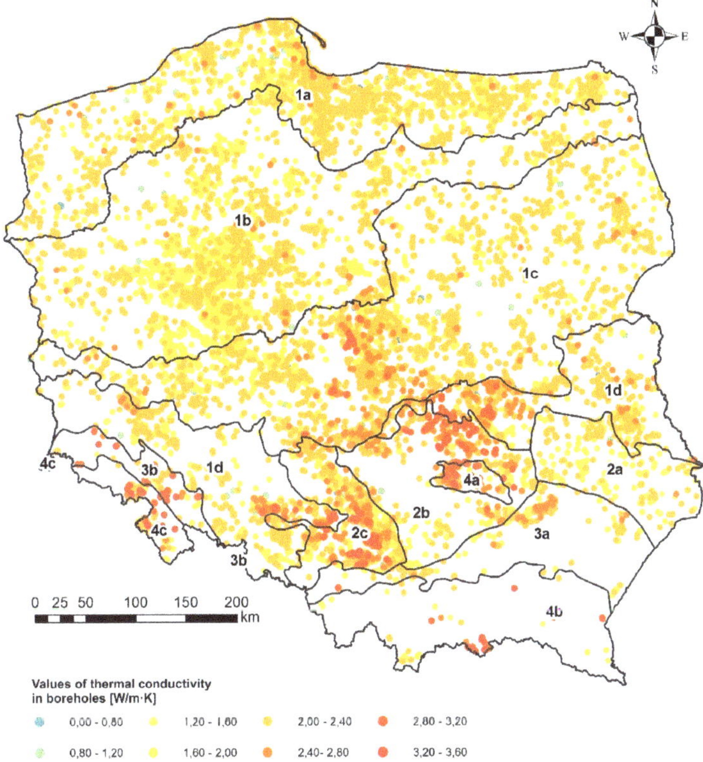

Figure 8. Map of the shallow geothermal potential of Poland by Ryżyński et al. [24], modified.

Regions highlighted on the map are: Pomorian phase of North Polish Glaciation (1a), Vistulian phase of North Polish Glaciation (1b), Wartanian phase of Middle Polish Glaciation (1c), Odranian phase of Middle Polish Glaciation (1d), Lubelska Upland (2a), Malopolska Upland (2b), Slasko-Krakowska Upland (2c), Carpathian Foothills (3a), Sudethian Foothills (3b), Holy-Cross Mountains (4a), Carpathian Mountains (4b), Sudethian Mountains (4c).

Regions from 1a to 1d are covered with a relatively thick layers of quaternary soils deposit: gravels, sands, silt and clays. The calculations and map preparation were possible due to the application of the presented methodology, which made it possible to obtain a large amount of data on the thermal properties of soils occurring in the above-mentioned geological regions.

5. Conclusions

The presented methodology was based on the functioning ISO and ASTM standards regarding geotechnical investigation and testing of physico-mechanical and thermal parameters of soil, which allows it to be easily implemented in geological and geotechnical laboratories. Sample preparation and measurements are relatively quick and cheap. The methodology provides repeatable and reliable results (consistent with archival data), thus it has been used in the process of preparation of serial shallow geothermal potential maps and parameterization of geothermal 3D models.

After performing a series of measurements, the authors consider it necessary to prepare databases of thermal properties of soils in a regional and lithogenetic approach, which will be the subject of future research.

Author Contributions: A.Ł.: resources, methodology, investigation, visualization, writing—original draft preparation. G.R.: conceptualization, supervision, validation, writing—review and editing. M.Ż.: methodology, investigation, data curation, formal analysis, writing—original draft preparation. All authors have read and agreed to the published version of the manuscript.

Funding: This research received no external funding.

Conflicts of Interest: The authors declare no conflict of interest.

References

1. Florides, G.A.; Pouloupatis, P.D.; Kalogirou, S.; Messaritis, V.; Panayides, I.; Zomeni, Z.; Partasides, G.; Lizides, A.; Sophocleous, E.; Koutsoumpas, K. The geothermal characteristics of the ground and the potential of using ground coupled heat pumps in Cyprus. *Energy* **2011**, *36*, 5027–5036. [CrossRef]
2. Ramos, R.; Aresti, L.; Christodoulides, P.; Vieira, A.; Florides, G. Assessment and Comparison of Soil Thermal Characteristics by Laboratory Measurements. In *Energy Geotechnics. SEG 2018*; Springer Series in Geomechanics and Geoengineering; Ferrari, A., Laloui, L., Eds.; Springer: Cham, Switzerland, 2018; pp. 155–162. Available online: https://link.springer.com/chapter/10.1007/978-3-319-99670-7_20 (accessed on 16 January 2019).
3. Vieira, A.; Alberdi-Pagola, M.; Christodoulides, P.; Javed, S.; Loveridge, F.; Nguyen, F.; Cecinato, F.; Maranha, J.; Florides, G.; Prodan, J.; et al. Characterisation of Ground Thermal and Thermo-Mechanical Behaviour for Shallow Geothermal Energy Applications. *Energies* **2017**, *10*, 2044. [CrossRef]
4. Ryżyński, G.; Bogusz, W. City-scale perspective for thermoactive structures in Warsaw. *Environ. Geotech.* **2016**, *3*, 280–290. [CrossRef]
5. Bogusz, W. Możliwość zastosowania fundamentów termoaktywnych w budownictwie mostowym. *Mosty* **2017**, *1*, 42–45. (In Polish)
6. Baralis, M.; Barla, M.; Bogusz, W.; Di Donna, A.; Ryżyński, G.; Żeruń, M. Geothermal potential of the NE extension Warsaw (Poland) metro tunnels. *Environ. Geotech.* **2018**, 1–13. Available online: https://www.icevirtuallibrary.com/doi/abs/10.1680/jenge.18.00042 (accessed on 16 January 2019). [CrossRef]
7. Anders, G.J.; Radhakrishna, H.S. Power cable thermal analysis with consideration of heat and moisture transfer in the soil. *IEEE Trans. Power Deliv.* **1988**, *3*, 1280–1288. [CrossRef]
8. Popov, Y.; Beardsmore, G.; Clauser, C.; Roy, S. ISRM suggested methods for determining thermal properties of rocks from laboratory tests at atmospheric pressure. *Rock Mech. Rock Eng.* **2016**, *49*, 4179–4207. [CrossRef]
9. Ryżyński, G.; Łukawska, A.; Żeruń, M. Wytyczne do prowadzenia seryjnych oznaczeń laboratoryjnych przewodności cieplnej gruntów spoistych i niespoistych na potrzeby sporządzania map i baz danych potencjału geotermii niskotemperaturowej. *Tech. Poszuk. Geol. Geoterm. Zrównoważony Rozw.* **2018**, *57*, 229–230. (In Polish)
10. *EN 1997-2:2007 Eurocode 7: Geotechnical Design—Part 2: Ground Investigation and Testing*; European Committee for Standardization: Brussels, Belgium, 2007; Available online: https://eurocodes.jrc.ec.europa.eu/showpage.php?id=137 (accessed on 5 December 2019).
11. ASTM. *ASTM D698-12e2, Standard Test Methods for Laboratory Compaction Characteristics of Soil Using Standard Effort (12 400 ft-lbf/ft3 (600 kN-m/m3))*; ASTM International: West Conshohocken, PA, USA, 2012.

12. ISO. *ISO 14688-2:2018 Geotechnical Investigation and Testing. Identification and Classification of Soil. Principles for a Classification*; International Organization for Standardization: Geneva, Switzerland, 2018; Available online: https://www.iso.org/ (accessed on 25 December 2019).
13. ISO. *ISO 17892-2:2014 Geotechnical Investigation and Testing. Laboratory Testing of Soil. Determination of Bulk Density*; International Organization for Standardization: Geneva, Switzerland, 2014; Available online: https://www.iso.org/ (accessed on 10 December 2019).
14. ISO. *ISO 17892-4:2017 Geotechnical Investigation and Testing—Laboratory Testing of Soil—Part 4: Determination of Particle Size Distribution*; International Organization for Standardization: Geneva, Switzerland, 2017; Available online: https://www.iso.org/ (accessed on 10 November 2019).
15. ISO. *ISO 17892-12:2018 Geotechnical Investigation and Testing—Laboratory Testing of Soil—Part 12: Determination of Liquid and Plastic Limits*; International Organization for Standardization: Geneva, Switzerland, 2018; Available online: https://www.iso.org/ (accessed on 18 November 2019).
16. ISO. *ISO 14688-1:2017 Geotechnical Investigation and Testing—Identification and Classification of Soil—Part 1: Identification and Description*; International Organization for Standardization: Geneva, Switzerland, 2017; Available online: https://www.iso.org/ (accessed on 21 November 2019).
17. ISO. *ISO 17892-1:2014 Geotechnical Investigation and Testing. Laboratory Testing of Soil. Determination of Water Content*; International Organization for Standardization: Geneva, Switzerland, 2014; Available online: https://www.iso.org/ (accessed on 5 December 2019).
18. *PN-B-04481:1988 Grunty Budowlane—Badania Próbek Gruntu (Geotechnical Soils—Soil Testing)*; The Polish Committee for Standardization: Warszawa, Polska, 1988; Available online: https://www.pkn.pl/en/ (accessed on 9 December 2019). (In Polish)
19. ASTM. *ASTM D5334-14 Standard Test Method for Determination of Thermal Conductivity of Soil and Soft Rock by Thermal Needle Probe Procedure*; ASTM International: West Conshohocken, PA, USA, 2014.
20. IEEE. *IEEE 442-2017—IEEE Guide for Thermal Resistivity Measurements of Soils and Backfill Materials*; The Institute of Electrical and Electronics Engineers: New York, NY, USA, 2017; Available online: https://www.ieee.org/ (accessed on 12 November 2019).
21. Baza Danych Geologiczno-Inżynierskich BDGI. Available online: http://geoportal.pgi.gov.pl/atlasy_gi (accessed on 23 December 2019). (In Polish)
22. Frankowski, Z.; Majer, E.; Sokołowska, M.; Ryżyński, G.; Ostrowski, S.; Majer, K. Badania geologiczno-inżynierskie prowadzone w Państwowym Instytucie Geologicznym w drugim pięćdziesięcioleciu jego działalności. *Prz. Geol.* **2018**, *66*, 752–768. (In Polish)
23. Majer, E.; Sokołowska, M.; Frankowski, Z.; Barański, M.; Bestyński, Z.; Ostrowski, S.; Pasieczna, A.; Pietrzykowski, P.; Przyłucka, M.; Błachnio, O.; et al. *Zasady Dokumentowania Geologiczno-Inżynierskiego*; Dział Wydawnictw PIG-PIB: Warsaw, Poland, 2018; pp. 34–35. (In Polish)
24. Ryżyński, G.; Żeruń, M.; Kocyła, J.; Kłonowski, M.R. Estimation of Potential Low-temperature Geothermal Energy Extraction from the Closed loop Systems Based on Analysis, Interpretation and Reclassification of Geological Borehole Data in Poland. In Proceedings of the World Geothermal Congress 2020, Reykjavik, Iceland, 26 April—2 May 2020.
25. Farouki, O.T. *Thermal Properties of Soils*; Cold Regions Research and Engineering Laboratory Monograph 81-1: New Hampshire, MA, USA, 1981.
26. Robertson, E.C. *Thermal Properties of Rocks*; Open-File Report no. 88-441; US Geological Survey: Richmond, VA, USA, 1988. Available online: https://pubs.er.usgs.gov/publication/ofr88441 (accessed on 12 January 2019).
27. Maglić, K.D.; Cezairliyan, A.; Peletsky, V.E. (Eds.) *Compendium of Thermophysical Property Measurement Methods: Volume 2 Recommended Measurement Techniques and Practices*; Springer: Boston, MA, USA, 1992.
28. Carslaw, H.S.; Jaeger, J.C. *Conduction of Heat in Solids*; Oxford Science Publications: Oxford, UK, 1959.
29. *KD2 Pro Thermal Properties Analyzer Operator's Manual Version 4*; Decagon Devices: Pullman, DC, USA, 2016; Available online: https://www.metergroup.com/ (accessed on 18 January 2019).
30. Myślińska, E. *Laboratoryjne Badania Gruntów*; Wydawnictwo Naukowe PWN: Warsaw, Poland, 2006. (In Polish)
31. *Thermophysical Instruments—Geothermal Investigation TK04 Thermal Conductivity Meter User's Manual version 5.5*; Thermophysical Instruments—Geothermal Investigation: Berlin, Germany, 2019; Available online: https://www.te-ka.de/en/ (accessed on 16 January 2019).

32. Polska Organizacja Rozwoju Technologii Pomp Ciepła. *Wytyczne Projektowania, Wykonania i Odbioru Instalacji z Pompami Ciepła. Część 1—Dolne źródło Ciepła*; Wydanie 1: Krakow, Poland, 2013. (In Polish)
33. VDI 4640 Blatt 2 2001 Thermal use of the underground—Ground source heat pump systems. 2001.
34. Bolcon Sweden. *EED 3.0-Earth Energy Designer*; Bolcon AB: Lund, Sweden, 2008; Available online: https://buildingphysics.com (accessed on 29 November 2019).
35. Santaa, G.D.; Galgaroa, A.; Sassia, R.; Cultreraa, M.; Scottona, P.; Muellerc, J.; Bertermannc, D.; Mendrinosd, D.; Pasqualie, R.; Peregoa, R.; et al. An updated ground thermal properties database for GSHP applications. *Geothermics* **2020**, *85*, 101758. [CrossRef]

Article

Investigation of Steady-State Heat Extraction Rates for Different Borehole Heat Exchanger Configurations from the Aspect of Implementation of New TurboCollector™ Pipe System Design

Tomislav Kurevija [1,*], Adib Kalantar [2], Marija Macenić [1] and Josipa Hranić [1]

1 Faculty of Mining, Geology and Petroleum Engineering, University of Zagreb, 10000 Zagreb, Croatia; mmacenic@rgn.hr (M.M.); josipa.hranic@rgn.hr (J.H.)

2 Faculty of Textiles, Engineering and Business, University of Borås, SE-501 90 Borås, Sweden; adib.kalantar@hb.se

* Correspondence: tkurevi@rgn.hr; Tel.: +385-1-553-5834

Received: 26 February 2019; Accepted: 18 April 2019; Published: 20 April 2019

Abstract: When considering implementation of shallow geothermal energy as a renewable source for heating and cooling of buildings, special care should be taken in the hydraulic design of the borehole heat exchanger system. Laminar flow can occur in pipes due to the usage of glycol mixtures at low temperature or inadequate flow rates. This can lead to lower heat extraction and rejection rates of the exchanger because of higher thermal resistance. Furthermore, by increasing the flow rate to achieve turbulent flow and satisfactory heat transfer rate can lead to an increase in the pressure drop of the system and oversizing of the circulation pump which leads to impairment of the seasonal coefficient of performance at the heat pump. The most frequently used borehole heat exchanger system in Europe is a double-loop pipe system with a smooth inner wall. Lately, development is focused on the implementation of a different configuration as well as with ribbed inner walls which ensures turbulent flow in the system, even at lower flow rates. At a location in Zagreb, standard and extended thermal response tests were conducted on three different heat exchanger configurations in the same geological environment. With a standard TRT test, thermogeological properties of the ground and thermal resistance of the borehole were determined for each smooth or turbulator pipe configuration. On the other hand, extended Steady-State Thermal Response Step Test (TRST) incorporates a series of power steps to determine borehole extraction rates at the defined steady-state heat transfer conditions of 0/−3 °C. When comparing most common exchanger, 2U-loop D32 smooth pipe, with novel 1U-loop D45 ribbed pipe, an increase in heat extraction of 6.5% can be observed. Also, when the same comparison is made with novel 2U-loop D32 ribbed pipe, an increase of 18.7% is achieved. Overall results show that heat exchangers with ribbed inner pipe wall have advantages over classic double-loop smooth pipe designs, in terms of greater steady-state heat extraction rate and more favorable hydraulic conditions.

Keywords: shallow geothermal; borehole heat exchanger; heat pump; renewable energy; applied thermogeology

---

1. Introduction and Literature Overview

The interest of using shallow geothermal energy via borehole heat exchangers and heat pump systems is on the rise in the last decades. In order to optimize the system and to determine its performance, thermal response tests are usually performed on heat exchangers. The most common pipe configurations installed in boreholes are double-U (2U) or single-U (1U) ones. Therefore, these configurations, with different diameters and inside pipe lining, were chosen to compare their thermogeological and hydraulic parameters and the impact they have on heat rejection/extraction rates.

The thermal response test (TRT) is a standard in-situ method of evaluating thermogeological properties of the ground, especially the effective ground thermal conductivity coefficient. The procedure consists of circulating a heated fluid through a borehole heat exchanger (BHE), which causes heat rejection to the ground. With the inversion of collected data, the temperature response in the case of heat extraction could be obtained. With such an analysis, the optimization of borehole heat exchanger fields and heat pump systems is possible. Even though there is already extensive research on the implementation of TRT data on optimizing BHE systems, there is space to improve and implement the research, especially when it comes to hydraulic setting.

In the last decade alone there has been a rise of research when it comes to optimizing the BHE system, by observing the influence of various geometrical settings, hydraulic settings and heat rates on the overall performance. Various studies have focused on a better understanding and more precise estimation of effective thermal conductivity and borehole thermal resistance [1–4] as important parameters when exploiting the shallow geothermal resource. Current methods of determining the average fluid temperature within BHE, as well as the influence of various measuring methods during TRT, on the final determination of thermal properties are also available [5–7]. With newly established data interpretation of time derivative of fluid temperature, measured during the first few hours of TRT operation, it was shown that the duration of the test could be shortened significantly, while the values of effective thermal conductivity can still be estimated within the 10% margin of error [8]. A similar result of only a 10% value variation of thermal conductivity, can be obtained if the data are obtained from the falloff temperature decline when using a method based on the analogy between TRT and petroleum well testing [9]. The origin of both procedures lies in the same diffusivity differential equation, with solutions for heat conduction or pressure transient analysis during radial flow in porous media. The use of the step thermal response test showed that it is suitable when determining heat rejection/extraction rates, useful for design optimization of the BHE field [3,10]. The influence of the groundwater advection on the efficiency and modelling of the ground BHE is detailed, with the study of its relationship with the modeling of the BHE field [7,11–13]. Aside from the borehole distance and the geometrical arrangement of the borehole field, research was done on the comparison of a geometrical setting of the exchanger pipes themselves.

Some recent studies deal with a comparison of various borehole heat exchanger types, constructions and overall efficiency of the ground source heat pump systems [1,14–16]. When comparing single U and double U tubes, it was shown that the double U exchangers show better performance, with reduced thermal resistance values [17]. There is also a development of various other BHEs setup, other than classic 1U and 2U; such are coaxial, helical and oval pipes and the use of the so-called fins or grooves within the pipe itself [1,18–20]. Among the first to study the use of the ribbed inner wall or micro fins, was Acuña [21]. Fins enable higher heat transfer rates due to the larger heat transfer surface, which improves the thermal performance of the exchanger pipe. Additionally, the research showed lower pressure drop values, in comparison with other tested BHEs. Also, the function of such a configuration is to maintain turbulent fluid flow, even in situations where the Reynolds number suggests a laminar type fluid flow. This results in lower use of the circulating pump, which in turn affects the operating costs. Since then, models and experiments have been presented on the benefits of using finned pipes, commercially known as TurboCollector^{TM} [22].

Bouhacina et al. [23] presented two numerical models, which described heat transfer and showed effects when using finned high-density polyethylene (HDPE) 1U pipe. The results were compared with a classic 1U layout with a smooth inner wall pipe. It was concluded that because of the smaller hydraulic diameter the mass flow rate is lower with finned pipes than in smooth pipes. Bae et al. [24] conducted TR tests at the same site on four different BHEs-coaxial, 1U HDPE, 1U HDPE-nano and 1U finned pipes. The results showed that with the improvement of convective heat transfer coefficient and thermal conductivity of pipe, there is a positive influence on overall heat transfer. The borehole thermal resistance of finned pipe was lower than the value of standard HDPE pipe, which was attributed to the fact that the pipe and fluid thermal resistances are reduced.

Therefore, the objective of the paper is to analyze the difference in steady-state heat extraction and rejection rates between different pipe types and geometry of the borehole heat exchangers in the same geological environment in a real project location. The main postulate of the research is the axiom that the novel TurboCollector™ exchangers, with a ribbed inner wall, have a greater heat extraction and rejection rates due to lower thermal resistance or skin, because of the higher convective heat transfer between fluid and pipe wall. Furthermore, TurboCollector™ exchangers should have lower pressure drops when turbulent regime is achieved, due to the lower required flow rate in pipes. This hypothesis is crucial to promote this novel design in the case of deeper drilling, where borehole heat exchangers work with more favorable ground temperature due to the positive effect of the geothermal gradient.

2. Theoretical Background

The mathematical model, which describes the extraction or rejection of the heat in the underground, is based on the heat diffusivity equation. Fourier [25] established the model, known as Fourier's law, and it describes conductive heat transfer in the homogeneous and isotropic environment. When using the cylindrical coordinates, it is expressed as:

$$\frac{1}{r}\frac{\partial}{\partial r}\left(r\frac{\partial T}{\partial r}\right) = \frac{\lambda}{\rho c}\frac{\partial T}{\partial t}, \tag{1}$$

where T is temperature (°C), r is radius (m), λ is thermal conductivity of the material (W/m °C), ρ is material density (kg/m^3), c is specific heat (J/kg °C) and t is time (s). The expression $\lambda/\rho c$ Kelvin described as thermal diffusivity, α(m^2/s). Ingersoll and Zobel [26] describe the Fourier's law of heat conduction, in solids, with a partial differential equation. Using cylindrical coordinates equation becomes:

$$\alpha\left[\frac{\partial^2 T}{\partial r^2} + \frac{1}{r}\frac{\partial T}{\partial r} + \frac{1}{r^2}\frac{\partial^2 T}{\partial \phi^2} + \frac{\partial^2 T}{\partial z^2}\right] = \frac{\partial T}{\partial t}. \tag{2}$$

There are two main analytical models when solving Equation (2)—line source and cylindrical model. The solutions are dependent on boundary conditions which are taken into account for each of the model. However, in both models, the term $\frac{1}{r^2}\frac{\partial^2 T}{\partial \phi^2}$ is neglected [27]. There are two main approaches when solving for line source model—the infinite line source and finite line source model. The infinite line source model, first described and solved by Lord Kelvin, describes radial heat flow. Carslaw and Jaeger [28] gave a well-known solution for an infinite line source model, used for describing heat extraction or rejection when using borehole heat exchangers. The model assumes that the borehole is an infinite line source in a homogenous and isotropic medium. The solution describes the temperature distribution as a function of time at some distance in relation to the borehole. Since the vertical component is neglected ($\partial^2 T/\partial z^2 = 0$) only radial heat flow is observed. The analytical solution includes the use of exponential integral or its simplified form, with certain constrictions [29–32]. In the case of heat rejection, the expression for determining temperature response, while performing the thermal response test is:

$$T(r_b, t) = T_0 + \frac{q'}{4\pi\lambda}\left(Ei\left(\frac{r_b^2}{4\alpha t}\right)\right), \tag{3}$$

where $T\ (r_b,\ t)$ is the temperature in function of radius and time (°C), T_0 represents the undisturbed ground temperature or the initial temperature (°C), q' heat power per meter of a borehole (W/m), r_b borehole radius (m) and Ei represents the exponential integral, which can be approximated with natural logarithmic function, in cases when $(4\alpha t/r^2) > 50$:

$$-E_i(-x) \cong -\ln(\gamma x) \cong -\ln(e^{\gamma} x) = \ln\left(\frac{1}{x}\right) - \gamma = \ln\left(\frac{1}{x}\right) - 0.5772, \tag{4}$$

where γ is Euler's constant with the value of 0.5772. Then, the ultimate expression for heat rejection is:

$$T(r_b,t) = T_0 - \frac{q'}{4\pi\lambda}\left(\ln\left(\frac{e^{\gamma}r_b^2}{4\alpha t}\right)\right) = T_0 - \frac{q'}{2\pi\lambda}\left(\ln\frac{r_b^2}{4\alpha t} + 0.5772\right) = T_0 - \frac{q'}{4\pi\lambda}\left(\ln\frac{r_b^2}{\alpha t} - 0.80907\right), \quad (5)$$

where $\frac{q'}{4\pi\lambda} = \kappa$ is the slope of the line when plotting T vs. $\ln(t)$ and it is a standardized principle to obtain average ground thermal conductivity [32].

The finite line source (FLS) model considers the vertical component ($\partial^2 T/\partial z^2 \neq 0$) as one of the boundary conditions, i.e., the finite length of the heat exchanger is considered. Claesson and Eskilson [27,29,33] gave the solution for the temperature at the borehole wall with FLS model, and in the case of heat rejection it is:

$$T(r_b,t) = T_0 + q'\frac{1}{2\pi\lambda}\,g\left(\frac{t}{t_s},\frac{r_b}{H}\right) \quad (6)$$

where H is the borehole depth (m) and g represents the so-called g-function (or Eskilson's g-function) and it is described as dimensionless temperature response factor. The g-function is calculated as:

$$g\left(\frac{t}{t_s},\frac{r}{H}\right) = \begin{cases} \ln\left(\frac{H}{2r}\right), & t > t_s \\ \ln\left(\frac{H}{2r}\right) + \frac{1}{2}\ln\left(\frac{t}{t_s}\right), & \frac{5r^2}{\alpha} < t < t_s \end{cases} \quad (7)$$

The influence of thermal properties of the soil is expressed through the factor t_s (s), which represents the time at which the steady state heat flow is achieved:

$$t_s = \frac{H^2}{9\alpha}. \quad (8)$$

In the denominator, this equation contains a value of thermal diffusivity, α, which is assumed from the drilling data. A compressed air system was used to lift the drill cuttings from the well bottom. Based on observed cuttings, the lithological profile was constructed. Based on the catalogue values for a specific type of soil from the lithological profile, ground thermal diffusivity coefficient was estimated. Since this value is usually estimated in practical projects, the method of determining the duration of the transition period can cause a further error in interpretation, especially for highly heterogeneous ground. Much more accurate graph-analytical method, the so-called derivation curve principle, can be implemented to determine the transition from unsteady state heat flow to relevant semi-steady state heat flow regime [9]. Derivation curve constructed from the obtained borehole fluid temperature (y) and TRT elapsed time (x) is derived as:

$$\left(\frac{dy}{dx}\right)_A = \frac{\left(\frac{y_1}{x_1}x_2 + \frac{y_2}{x_2}x_1\right)}{(x_1 + x_2)}. \quad (9)$$

In order to predict the fluid temperature inside the borehole heat exchanger, the value of borehole thermal resistance must be determined. Well testing in petroleum engineering, when defining formation permeability, is analogous to the classic TRT data analysis when defining thermal conductivity. Considering the similarities in the origin of the equations describing the behavior of pressure in porous medium and heat conduction in solids [9,10], borehole thermal resistance is also equivalent to near-well damage formation or skin factor, s, which affects the fluid flow in the well. In thermogeology, the skin factor describes the dimensionless thermal resistance to the heat flow in the borehole.

It depends mainly on pipe configuration, the grout used for cementing of the borehole, as well as on the fluid type and flow properties, and it is expressed as:

$$s = \frac{1}{2}\ln\left(\frac{e^{\gamma} r_b^2}{4\alpha t_p}\right) - \frac{\left(T_0 - T_p\right)2\pi\lambda}{q'}, \tag{10}$$

where t_p is the duration of power step during semi-steady state and $T_0 - T_p$ is a temperature difference between initial temperature and temperature at the end of the power step. The analogy between dimensionless skin factor and thermal resistance is expressed as initial temperature rise due to heat flow in the pipe wall and grout, ΔT_{skin} [9,30,31]:

$$\Delta T_{skin} = s\left(\frac{q'}{2\pi\lambda}\right) = R_b' \cdot q', \tag{11}$$

where R' is the equivalent borehole thermal resistance (°C m/W). A more precise estimation of the thermal conductivity and borehole skin factor is crucial when defining borehole exchanger heat power capacity during long-term operation. In our previous research [9], we showed that implementing the novel steady-state thermal response step test (TRST) is helpful with system optimization. This is especially important for systems with long annual full-load hours of operation when there is a need to establish a relationship between peak load working conditions of the heat pump system and steady-state entering source temperature from the bore field (EST). Also, the latest development of inverter-type geothermal heat pumps requires such an analysis, since the inverter compressor works continuously with modulating power, depending on the outside air temperature curve [34].

The step test is carried out by imposing different heat pulses (heat rejection rates) after a certain period. By using Eskilson's g-function analysis, it is possible to analytically describe the temperature response of conducted TRST for each step, by superimposing each following step on the steps already conducted [27,29,33]. Therefore, for any arbitrary heat pulse, in case of TRT-heat rejection pulse, the thermal response can be found. The superposition principle suggests that, in the case of performing TRT with three different pulses, the first thermal pulse is imposed during the entire testing period. Each following pulse is superimposed on the previous one. The general expression for the average fluid temperature inside the BHE, for different heat rejection [27] is:

$$T_f(t) = T_0 + \left[\frac{1}{2\pi\lambda}\sum_{i=1}^{n}\left(q_i' - q_{i-1}'\right)g_i\right] + \Delta T_{skin}, \tag{12}$$

where $T_f(t)$ is borehole fluid mean temperature (°C), q_i' and q_{i-1}' are consequent heat power steps. For example, in the case of three different heat rejection pulses the average temperature of the circulating fluid would be calculated as follows:

First step:

$$T_f(t) = T_0 + \left[\left(\frac{q_1'}{2\pi\lambda}\right) \cdot g_{t_1}\right] + \Delta T_{skin1}, \tag{13}$$

Second step:

$$T_f(t) = T_0 + \left(\frac{1}{2\pi\lambda}\right)\left[\left(q_1' \cdot g_{t_1}\right) - \left(-q_2' + q_1'\right) \cdot g_{t_2 - t_1}\right] + \Delta T_{skin2}, \tag{14}$$

Third step:

$$T_f(t) = T_0 + \left(\frac{1}{2\pi\lambda}\right)\left[\left(q_1' \cdot g_{t_1}\right) - \left(\left(-q_2' + q_1'\right) \cdot g_{t_2 - t_1}\right) - \left(\left(-q_3' + q_2'\right) \cdot g_{t_3 - t_2}\right)\right] + \Delta T_{skin3}. \tag{15}$$

3. Experimental Site Setting

3.1. Thermogeological and Hydrogeological Setting

The thermal response test was conducted in the city of Zagreb, Croatia, at the location as shown in Figure 1. It also shows a detailed geological map of the city of Zagreb and its surrounding area [35]. The area is located mostly on the Zagreb aquifer system, which is from the Quaternary Age. The aquifer is mostly comprised of Middle and Upper Pleistocene, and Holocene sediments. The location of the BHEs is near the Zagreb aquifer boundary. Up to 110 m, the underground is mostly comprised of gray and brown clay, with some gravel layer at the shallow depth (Figure 2). It is seen from the lithology description, that there is a saturated layer of soil, with a thickness of around 5 m. The project location is near the northern aquifer boundary. From the map of hydraulic conductivity (Figure 2) it can be assessed that the values of hydraulic conductivity at the project location are around 0.3 cm/s. Considering that the Zagreb aquifer is of alluvial origin, and connected to the Sava river water table, the relatively thin layer of saturated gravel and sand at the project site has an insignificant effect on overall heat exchanger capacity.

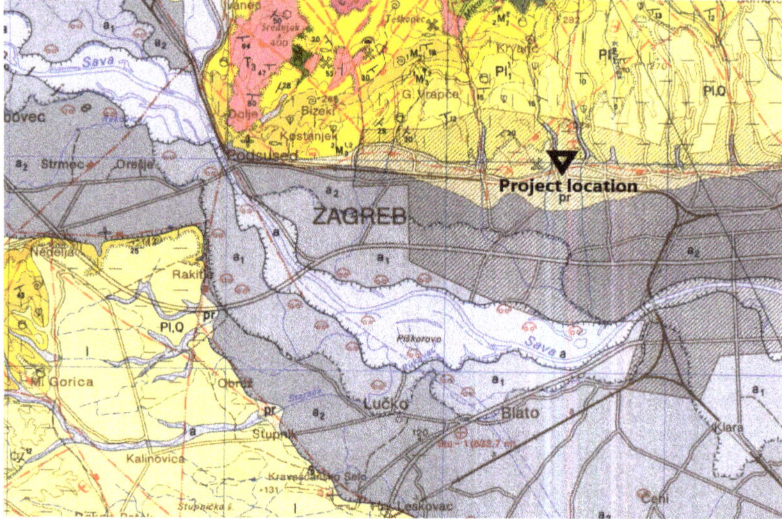

Figure 1. Detailed geological map with BHEs location and general lithology column of the project site. Legend for geology map: **a**—aluvium: gravels, sands and clays; $\mathbf{a_1}$—the lowest terrace: gravels, sands and clays to a lesser extent; $\mathbf{a_2}$—middle terrace: gravels and sands; **pr**—proluvium: gravels, sands and clays; **l**—clayey silt; **lb**—marshy loes: silty clays; **Pl,Q**—gravels, sands and clays; $\mathbf{Pl_1{}^1}$—marls, marly clays, sands to a lesser extent, sandstones, gravels and conglomerates (lower pont); $\mathbf{{}_2M_3{}^{1,2}}$—lime marls, sands to a lesser extent, sandstones, gravels and conglomerates (upper panon); $\mathbf{{}_2M_2{}^2}$—limestones, sandstones, lime and clayey marls (upper Tortonian).

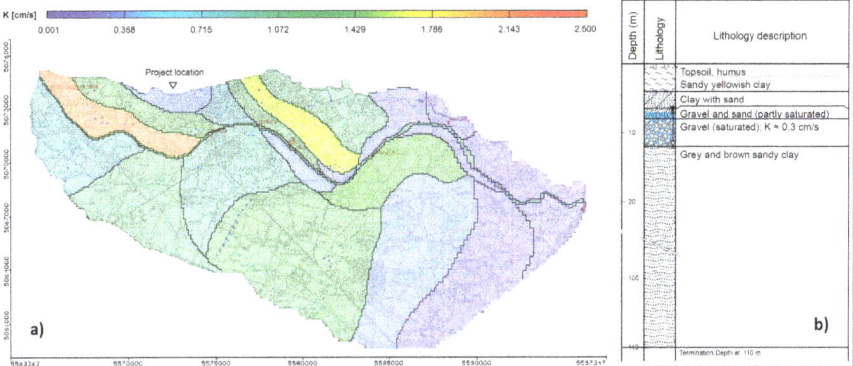

Figure 2. (**a**) Map of hydraulic conductivity of Zagreb aquifer [36] and (**b**) lithology of project location.

3.2. Borehole Heat Exchangers Setup at the Test Site

At the residential project site, three vertical boreholes for heating and cooling purposes were drilled up to 110 m. The drilling diameter of each of the borehole is 152 mm. The project site during the drilling and testing operations is seen in Figure 3a. The thermal response test was carried out on all of the three BHEs with the different geometrical setting, and two pipe arrangements.

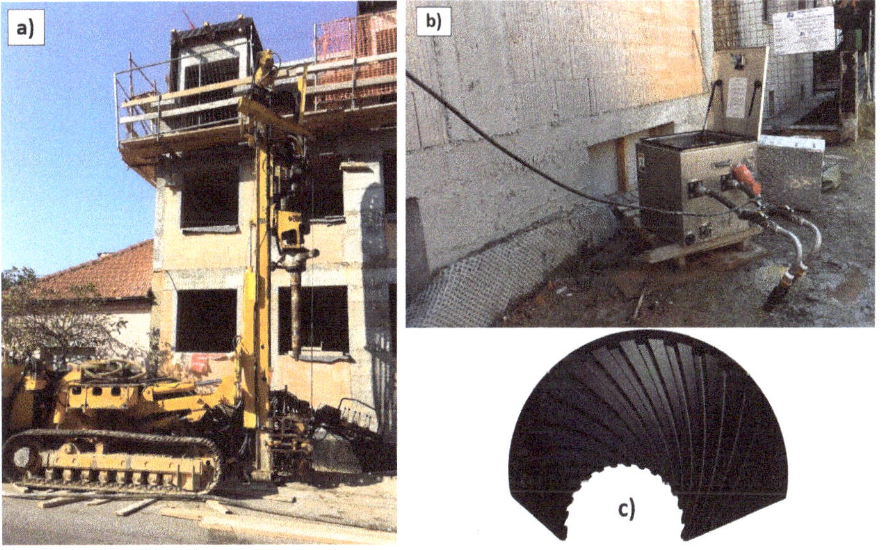

Figure 3. (**a**) Drilling of the BHEs; (**b**) performing the thermal response test; (**c**) TurboCollectorTM pipe wall.

The BHE-1 is equipped with polyethylene 2U-pipe (D32 mm PE100 SDR11) with a smooth inner wall. Such a design is the most often used borehole heat exchanger in Europe. The second heat exchanger, denoted as BHE-2, is equipped with 2U-pipe (D32 mm PE100 SDR11) with a ribbed inner wall, as shown in Figure 3c. The third borehole, BHE-3, was equipped with a novel TurboCollector^{TM} 1U-pipe (D45 mm PE100 SDR11).

The total measuring TRT time for each of the boreholes was approximately 120 hours. The three different pipe configurations were used in this project to determine the effect of borehole geometry on overall heating/cooling extraction/rejection rates. Boreholes are placed in typical L-shape, with separation between boreholes of 7 m, as shown in Figure 4.

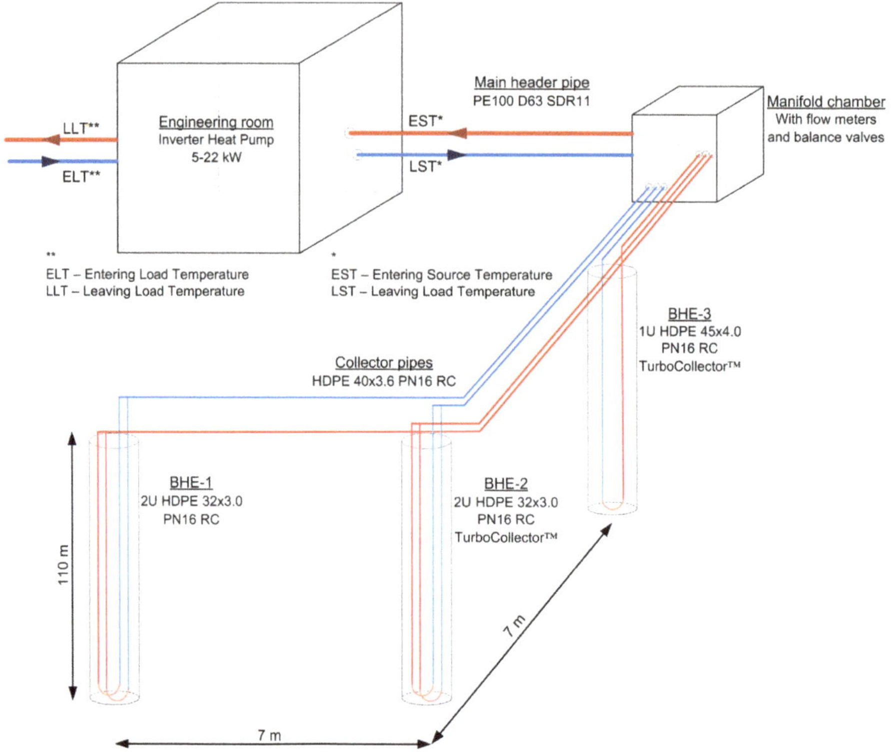

Figure 4. Simplified schematic of the borehole heat exchangers collector system.

Measurements on three borehole heat exchangers were conducted in October 2018 with a Geocube GC500 TRT apparatus (Precision Geothermal LLC, Marple Plain, MN, USA). The equipment has maximum available electric heater power of 9.0 kW @ 240 V. An internal logger (Onset HOBO H22 energy logger, Onset, Bourne, MA, USA) was set-up to collect inlet and outlet borehole temperature, flow rate, voltage and electric current data in 5-min intervals. Thermocouples (type: resistance temperature detectors-RTD) are placed on inlet and outlet of TRT and have an accuracy of ±0.2 °C ranging from 0 °C to 50 °C. The testing procedure was set up as a classic TRT heat rejection with an initial step of 6.5 kW and a duration of approximately 63 h. Two additional heat steps were introduced afterward, with lower rejection rates where stabilization of borehole temperature was observed. Second heat step of 4.5 kW lasted approximately 24 h, while the third heat step of 2.4 kW was conducted approximately 33 h.

The mean ground temperature along the borehole was measured with TRT apparatus at 15.2 °C, with the fluid flow of 0.45 L/s and no heaters switched on. Borehole fluid is comprised solely out of the water, with viscosity presumed to be around 1 mPa s.

4. Results and Discussion

After switching on the TRT heaters, the average temperature in the borehole heat exchanger begins to grow as a function of ground thermogeological parameters, power step, and borehole thermal resistance. Figure 5 shows recorded average fluid temperature data and power for the first heat step.

The entire test was conducted without any significant deviations in the power supply, which interprets highly representative temperature curves.

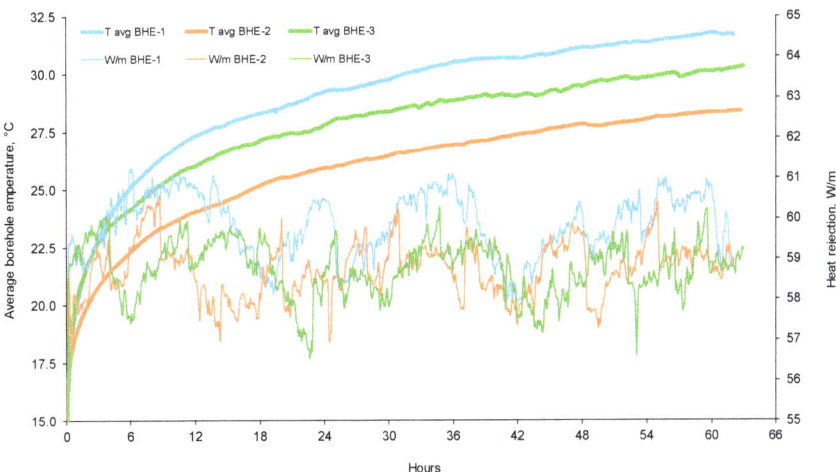

Figure 5. Recorded average fluid temperatures and a power step for a three borehole heat exchangers during TR testing.

To determine the effective ground thermal conductivity coefficient, it is necessary to plot the average temperature of circulating fluid data vs. natural logarithm of time, ln(t). According to the infinite line source method, initially recorded data during an unsteady state of heat transfer must be excluded from the analysis.

As seen from Figure 6a, using the derivative curve principle and Equation (9) it can be observed that all three BHE layouts shows a transition to semi-steady state heat transfer after roughly the 10th hour, where the derivative dT/dt falls below the value of 0.5 [6].

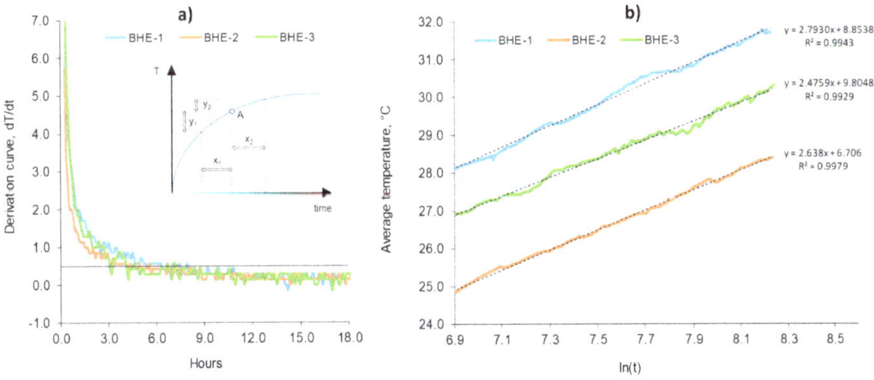

Figure 6. Determination of semi-steady state heat flow period (**a**) and determination of ground heat conductivity (**b**).

In Figure 6b the standard method is then applied to determine the effective ground thermal conductivity coefficient, by establishing slope of the straight line portion between average borehole temperature and the natural logarithm of time. Using Equation (5) the ground thermal conductivity is then calculated from all of the three tests (data presented in Table 1 and results in Table 2).

Table 1. Thermal Response Step Test obtained temperature data.

Heat Step	BHE-1 2U D32 Smooth	TRT Time	Heat Power	Heat Power Standard Deviation	Heat Power Standard Error	Cooling Cycle	Heating Cycle (Inversed)	Rejected Heat TRT
Step	Heat flow regime	h	W/m	W/m	-	EST °C	EST °C	kWh
1	unsteady	62.1	59.8	1.33	0.049	29.7	0.6	408.4
1a	semi-steady (FLS)	240.0	59.8	-	-	33.6	−3.2	-
2	steady state	23.9	40.9	0.35	0.021	27.0	3.4	107.7
3	steady state	32.7	21.2	0.23	0.011	22.9	7.5	76.5
4	initial conditions	0.0	0.0	-	-	15.2	15.2	0.0
	BHE-2 2U D32 Ribbed	**TRT Time**	**Heat Power**	**Heat Power Standard Deviation**	**Heat Power Standard Error**	**Cooling Cycle**	**Heating Cycle (Inversed)**	**Rejected Heat TRT**
Step	Heat flow regime	h	W/m	W/m	-	EST °C	EST inv. °C	kWh
1	unsteady	62.8	58.7	0.66	0.024	27.0	3.4	405.3
1a	semi-steady (FLS)	240.0	58.7	-	-	30.5	−0.1	-
2	steady state	24.0	40.8	0.54	0.032	25.0	5.4	107.8
3	steady state	33.4	21.5	0.28	0.014	21.8	8.6	79.2
4	initial conditions	0.0	0.0	-	-	15.2	15.2	0.0
	BHE-3 1U D45 Ribbed	**TRT Time**	**Heat Power**	**Heat Power Standard Deviation**	**Heat Power Standard Error**	**Cooling Cycle**	**Heating Cycle (Inversed)**	**Rejected Heat TRT**
Step	Heat flow regime	h	W/m	W/m	-	EST °C	EST inv. °C	kWh
1	unsteady	63.0	58.8	0.79	0.029	28.9	1.6	407.2
1a	semi-steady (FLS)	240	58.8	-	-	32.1	−1.7	-
2	steady state	24.5	40.0	0.42	0.025	26.1	4.3	107.9
3	steady state	33.4	20.7	0.22	0.011	22.2	8.2	76.2
4	initial conditions	0.0	0.0	-	-	15.2	15.2	0.0

Table 2. Thermal Response Step Test obtained temperature data.

Heat Step	BHE-1 2U D32 Smooth	Borehole Thermal Resistance	Skin Factor	ΔT Skin	Estimated λ 1st Step	Fitted λ Finite Line Source
Step	Heat flow regime	m °C/W	-	°C	W/m °C	W/m °C
1	unsteady	0.085	0.91	5.09	1.70	1.62
2	steady state	-	1.08	4.12	-	
3	steady state	-	1.48	2.94	-	
	BHE-2 2U D32 Ribbed	**Borehole Thermal Resistance**	**Skin Factor**	**ΔT Skin**	**Estimated λ 1st Step**	**Fitted λ Finite Line Source**
1	unsteady	0.039	0.43	2.28	1.77	1.81
2	steady state	-	0.57	2.11	-	
3	steady state	-	0.93	1.79	-	
	BHE-3 1U D45 Ribbed	**Borehole Thermal Resistance**	**Skin Factor**	**ΔT Skin**	**Estimated λ 1st Step**	**Fitted λ Finite Line Source**
1	unsteady	0.081	0.96	4.74	1.89	1.87
2	steady state	-	1.13	3.83	-	
3	steady state	-	1.53	2.68	-	

Figure 7 presents results of the extended thermal response testing. For each of the three boreholes additional fall-off tests were performed, by implementing additional lower heat power values until steady-state heat transfer was achieved for each of the steps (so-called Thermal Response Step Test—TRST). For each of the additional two heat steps, which lasted for 24 and 33 h, steady-state heat flow condition was observed. Such approach can provide confident information about dependency between two variables; borehole outlet fluid steady-state temperature in a function of heat pump evaporater/condenser peak load working conditions. Therefore, for any value of borehole heat rejection or extraction rate, stabilized fluid temperature in the borehole could be determined. This actually means that a borehole heat exchanger can work for a longer period without significant subsequent rise or drop in fluid temperature, depending on winter/summer regime.

Table 1 shows complete collected temperature data from Thermal Response Step Test with temperature stabilization values, in a function of heat rejection rate, as well as inversed temperature data for a theoretical case of heat extraction (mirrored temperatures in relation to initial temperature). Total of four steady-state points is then extracted from Figure 7 and presented in Table 1, including initial static temperature conditions. Table 2 shows obtained thermogelogical data as well as borehole resistance or skin factor.

The Finite Line Source (FLS) with Eskilson's g-function method was applied to the obtained temperature response data with three different heat rejection rates. The purpose was to create synthetic temperature response curves and compare them with real TRT obtained data. The calculation using Equations (12)–(15) was carried out with the results of the previously calculated thermogeological data, like ground thermal conductivity and borehole skin. The thermal diffusivity value in the complete analysis was predicted to be 0.060 m^2/d, which corresponds to catalogue data for moist clayey soils (Ground Loop Designer-GLD software). The results show a good match between temperature response from TRT and synthetic FLS temperature response, which can be seen in Figure 7. It has to be pointed out, that the g-function for designated periods (g_{t1}, g_{t2-t1}, g_{t3-t2}) in Equations (13)–(15) represent calculations for cumulative time for the intended steps. As explained in detail in our previous research [9] and repeated here, using a simple statistical analysis, such as the method of the Sum of Squares of Difference, could provide exactly which thermal conductivity coefficients are of statistical significance. Sum of Squares or Variation (SUMXMY2 function in MS Excel) is a statistical technique used in regression analysis to determine the dispersion of temperature points. In a regression analysis, the goal is to determine how well a data series (in this case measured entering source temperature-EST) can be fitted by a function which might help to explain how the data series, or temperature response, was generated (in this case synthetic temperature response constructed with FLS model and g-functions) [9]. The Sum of Squares method was used to find the function which best fits (varies least) from the real measured temperature with TRT. In Table 2 statistical results are presented in the way of solver solution, where function SUMXMY2 approaches zero for the best fit of the given ground thermal conductivity. It could be noticed that estimated conductivities from a TRT on three boreholes varies between 1.70–1.89 W/m °C, while conductivity obtained by a best fitting FLS solution with SUMXMY2 when approaching to zero value, is between 1.62–1.87 °C. This procedure gives high confidence in obtained results by TRT. Variation of thermal conductivity between boreholes is 10–15%, which can be explained by heterogeneity of the ground, as well as with daily power variations and heat losses in surface TRT equipment.

When observing calculated borehole thermal resistance from Table 2, it can be concluded that 2U D32 BHE-2 with TurboCollector^{TM} effect has a significantly lower value when compared to same geometrical 2U D32 BHE-1 but with the smooth pipe wall. This can be explained by the fact that the ribbed inside wall provides enhanced turbulent fluid flow regime, which then increases heat transfer flow between fluid and plastic pipe wall. This was also confirmed by other recent research related to borehole resistance evaluation between finned pipe and other pipe types [24].

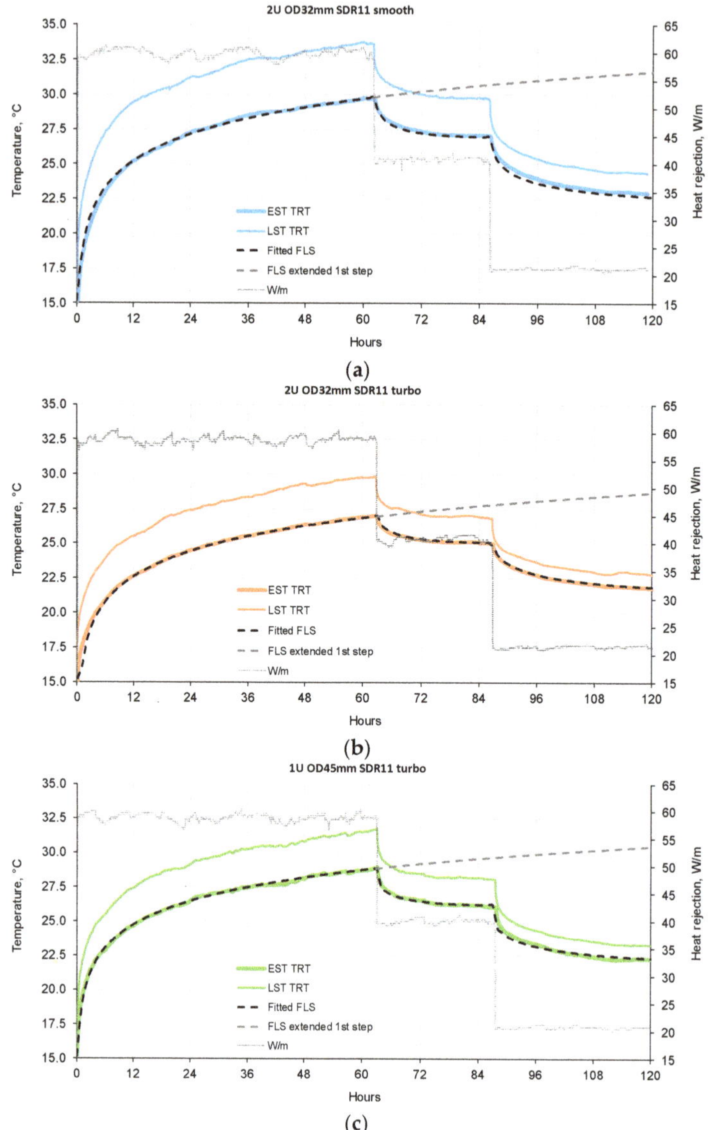

Figure 7. Fitting of measured data with FLS method (use of g-function) for three different pipe types: (**a**) 2U-pipe, smooth, D32 mm; (**b**) 2U-pipe, turbo, D32 mm; (**c**) 1U-pipe, turbo, D45 mm.

However, novel 1U D45 BHE-3 with TurboCollector^{TM} effect has similar borehole resistance as classic 2U D32 BHE-1. This can be explained by the fact that enhanced turbulent flow, due to the ribbed inner wall, offset thicker plastic pipe wall (4.0 mm compared to 3.0 mm for BHE-1).

As seen from Table 1, each of the performed heat steps is defined with its stabilized temperature, where steady-state heat transfer is achieved. As an additional zero power step, initial temperature conditions are introduced, in this case 15.2 °C, as an effective borehole temperature. By setting the steady-state temperature in each of the steps as separate points (Table 1), it is possible to construct the heat rejection and extraction diagram (W/m) as the function of the desired inlet temperature (entering source temperature—EST) to the heat pump, as seen from Figure 8.

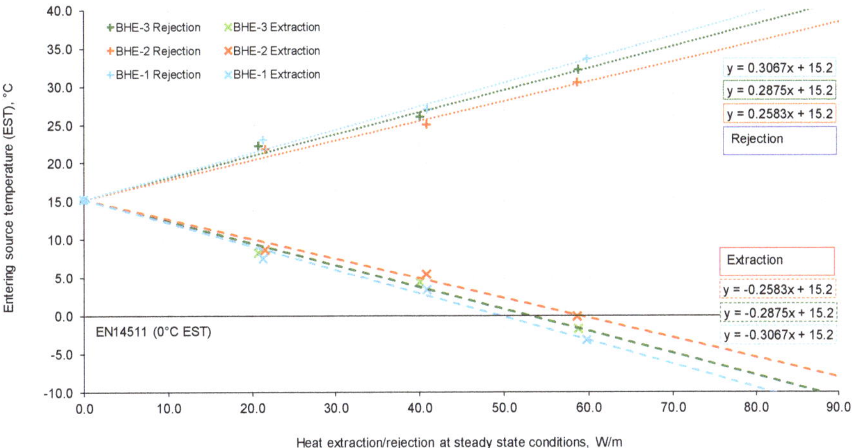

Figure 8. Determining the extraction/rejection heat for three different pipe configurations.

Defining exchanger heat power capacity in relation to steady-state heat transfer stabilization point could also secure the optimized selection of working fluid. Mixing a higher proportion of glycol with water gives confidence in the system design from the aspect of avoiding freezing problems, but it negatively influences fluid viscosity and fluid lower specific heat during heat extraction. Higher viscosity then negatively reflects on Reynolds number and the emersion of laminar flow in pipes, which raises borehole thermal resistance.

The fluid temperature in properly designed borehole heat exchangers (EST), which exploits shallow geothermal energy in combination with heat pumps, should never fall below 0 °C under peak load conditions. This value corresponds to the EN14511 testing standardized norm for a reliable coefficient of performance (COP) of the heat pump. At this value, the heat pump still efficiently provides heat energy compared to fossil fuel resources. Also, there is a minimum needed glycol mixture volume inside the borehole heat exchanger which assures better heat extraction rates due to turbulent flow in pipes. According to extraction/rejection steady-state equations from Figure 8 and EN14511 norm, borehole heat exchangers capacity is shown in Table 3.

Table 3. Obtained extraction and rejection heat rates from three BHE and from data presented in Figure 8.

Rejection/Extraction Capacity	BHE-1 2U D32 Smooth	BHE-2 2U D32 Ribbed	BHE-3 1U D45 Ribbed
-	W/m	W/m	W/m
Rejection capacity, according to EN14511 Heat pump COP = 5.10 (EST/LST 30/35 °C; LLT/ELT 7/12 °C)	64.7	76.7	68.9
Extraction capacity according to EN14511 Heat pump COP = 4.90 (EST/LST 0/−3 °C; LLT/ELT 35/30 °C)	49.6	58.9	52.9
Extraction capacity peak conditions Heat pump COP = 3.60 (EST/LST −5/−8 °C; LLT/ELT 35/30 °C)	65.9	78.3	70.3

Since in colder climate areas, where the undisturbed ground temperature per 100 m borehole is low (usually below 10 °C), the heat pump will often work with fluid temperatures below 0 °C, even during the base load. Therefore, Table 3 also shows heat extraction rates for fluid temperatures well below the water freezing point. Also, informative COPs of the inverter heat pump used in the project are presented (model ecoGEO B3 5-22kW, manufacturer: EU Spain-Ecoforest, Vigo, Spain), for all temperature regimes during heat rejection and extraction.

When observing BHE-1 (2U D32 smooth pipe), as most often heat exchanger installed in Europe, extraction capacity at site conditions is 49.6 W/m for EN14511 norm. Both other TurboCollector^{TM} heat exchangers show somewhat higher extraction capacity (58.9 W/m for 2U D32 ribbed and 52.9 W/m for 1U D45 ribbed), due to the more favorable turbulent flowing regime, which makes the implementation of such exchangers justifiable.

When comparing classic 2U D32 smooth pipe heat exchanger with the novel 1U D45 TurboCollector^{TM} pipe from a hydraulic point of view, it can be perceived from Figure 9 that pressure drop is significantly lower for 1U D45 exchanger for same flow conditions to the exchanger. Measured values of pressure drop for 1U D45 pipe were provided by the manufacturer (MuoviTech Brämhult, Sweden). These measurements were obtained as pressure difference by two sensitive manometers placed at the inlet and the outlet of the borehole heat exchanger. Pressure loss for a classic 2U D32 smooth pipe was calculated by the known empirical Darcy-Weisbach equation.

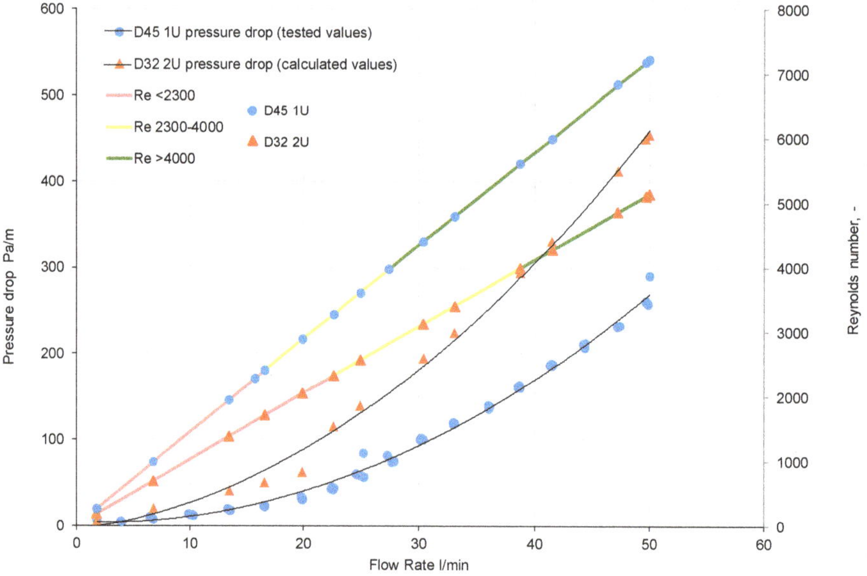

Figure 9. Hydraulic comparison of 2U D32 and 1U D45 TurboCollector^{TM} heat exchangers.

Furthermore, transitional and turbulent flow regime, defined by Reynolds number, occurs at lower flow rates for the 1U D45. This means that the circulation pump consumes less energy annually, giving a better seasonal coefficient of performance (SCOP) for the geothermal system.

5. Conclusions

To optimally design geothermal borehole heat exchanger grid it is crucial to understand thermogeological properties of the ground and borehole grout. Therefore, thermal response test is a vital procedure to determine ground thermal conductivity and borehole skin factor. Shallow geothermal systems with closed loops heat exchangers are often oversized or undersized in practice, due to uncertainty in ground properties or as the result of poor hydraulic design. In both cases, this leads to uneconomical geothermal project; unacceptable higher initial investment for oversized system and diminished efficiency of the heat pump for undersized system. Therefore, implementation of extended borehole thermal response step testing during project elaboration is cost-effective method to ensure longevity of the heat pump system and to predict borehole fluid temperature evolution during winter and summer months.

Investigation showed, that in the same similar geological environment, borehole heat exchangers comprised of TurboCollector™ ribbed inner wall offer somewhat higher heat extraction and rejection rates. When comparing standard, and most common system of smooth 2U D32 pipe exchanger, 1U D45 TurboCollector™ offers an increase in heat extraction by 6.5% and 2U D32 TurboCollector™ increase of 18.7%. This extraction rates comparison is strongly dependent on ground thermal conductivity, meaning that in lower thermal conductivity environment overall advantage in extraction rates between such heat exchangers would be smaller, and vice versa.

Also, implementation of 1U D45 TurboCollector™ heat exchanger should be forced when drilling deeper boreholes, due to significantly smaller pressure drops per meter of pipe. Deeper boreholes also provide a higher initial borehole temperature because of the geothermal gradient influence, and therefore higher extraction rates when considering the practical lower limit of 0 °C according to EN14511, or even lower. This is especially relevant for the northern hemisphere and higher latitudes, where initial ground temperature along the borehole can be rather low (usually 5–10 °C), so there is a significant amount of heat pump working hours below 0 °C. Even at borehole fluid temperature below 0 °C heat pump can still work relatively efficiently, so implementing deeper boreholes and ribbed exchangers can somewhat offset the initial loss in heat extraction when comparing for milder climate areas. Since the wider use of ribbed TurboCollector™ pipes in real projects is still a relatively new occurrence, there is a need for further research into their advantages over the current market dominant pipes with smooth inner walls.

Author Contributions: Conceptualization, T.K.; Data curation, T.K.; Formal analysis, J.H.; Investigation, T.K., M.M. and J.H.; Methodology, T.K. and M.M.; Resources, J.H.; Software, T.K. and M.M.; Supervision, A.K.; Validation, T.K. and A.K.; Visualization, M.M.; Writing—original draft, T.K.

Funding: This research received no external funding.

Conflicts of Interest: The authors declare no conflict of interest.

References

1. Zarrella, A.; Emmi, G.; Graci, S.; De Carli, M.; Cultrera, M.; Santa, G.D.; Galgaro, A.; Nertermann, D.; Müller, J.; Pockelé, L., et al. Thermal response testing results of different types of borehole heat exchangers: An analysis and comparison of interpretation methods. *Energies* **2017**, *10*, 801. [CrossRef]
2. Lamarche, L.; Raymond, J.; Koubikana Pambou, C.H. Evaluation of the internal and borehole resistances during thermal response tests and impact on ground heat exchanger design. *Energies* **2017**, *11*, 38. [CrossRef]
3. Boban, L.; Soldo, V.; Stošić, J.; Filipović, E.; Tremac, F. Ground Thermal Response and Recovery after Heat Injection: Experimental Investigation. *Trans. FAMENA* **2018**, *42*, 39–50. [CrossRef]
4. Fossa, M.; Rolando, D.; Pasquier, P. Pulsated Thermal Response Test experiments and modelling for ground thermal property estimation. In Proceedings of the IGSHPA Research Track, Stockholm, Sweden, 18–20 September 2018.
5. Beier, R.A.; Mitchell, M.S.; Spitler, J.D.; Javed, S. Validation of borehole heat exchanger models against multi-flow rate thermal response tests. *Geothermics* **2018**, *71*, 55–68. [CrossRef]
6. Beier, R.A. Use of temperature derivative to analyze thermal response tests on borehole heat exchangers. *Appl. Therm. Eng.* **2018**, *134*, 298–309. [CrossRef]
7. Badenes, B.; Mateo Pla, M.Á.; Lemus-Zúñiga, L.G.; Sáiz Mauleón, B.; Urchueguía, J.F. On the Influence of Operational and Control Parameters in Thermal Response Testing of Borehole Heat Exchangers. *Energies* **2017**, *10*, 1328. [CrossRef]

8. Pasquier, P. Interpretation of the first hours of a thermal response test using the time derivative of the temperature. *Appl. Energy* **2018**, *213*, 56–75. [CrossRef]
9. Kurevija, T.; Strpić, K.; Koščak-Kolin, S. Applying petroleum the pressure buildup well test procedure on thermal response test—A novel method for analyzing temperature recovery period. *Energies* **2018**, *11*, 366. [CrossRef]
10. Kurevija, T.; Macenić, M.; Strpić, K. Steady-state heat rejection rates for a coaxial borehole heat exchanger during passive and active cooling determined with the novel step thermal response test method. *Min. Geol. Pet. Eng. Bull.* **2018**, *33*, 61–71. [CrossRef]
11. Urchueguía, J.; Lemus-Zúñiga, L.G.; Oliver-Villanueva, J.V.; Badenes, B.; Pla, M.; Cuevas, J. How Reliable Are Standard Thermal Response Tests? An Assessment Based on Long-Term Thermal Response Tests Under Different Operational Conditions. *Energies* **2018**, *11*, 3347. [CrossRef]
12. Liuzzo-Scorpo, A.; Nordell, B.; Gehlin, S. Influence of regional groundwater flow on ground temperature around heat extraction boreholes. *Geothermics* **2015**, *56*, 119–127. [CrossRef]
13. Emad Dehkordi, S.; Schincariol, R.A.; Olofsson, B. Impact of groundwater flow and energy load on multiple borehole heat exchangers. *Groundwater* **2015**, *53*, 558–571. [CrossRef]
14. Tinti, F.; Giambastiani, B.; Mastrocicco, M. Types of Geo-Exchanger Systems for Underground Heat Extraction. In *Energy Science & Tecnology Vol. 9: Geothermal and Ocean Environment*; Sharma, U.C., Prasad, R., Sivakumar, S., Eds.; Studium Press: New Delhi, India, 2014.
15. Aresti, L.; Christodoulides, P.; Florides, G. A review of the design aspects of ground heat exchangers. *Renew. Sustain. Energy Rev.* **2018**, *92*, 757–773. [CrossRef]
16. Sarbu, I.; Sebarchievici, C. General review of ground-source heat pump systems for heating and cooling of buildings. *Energy Build.* **2014**, *70*, 441–454. [CrossRef]
17. Zeng, H.; Diao, N.; Fang, Z. Heat transfer analysis of boreholes in vertical ground heat exchangers. *Int. J. Heat Mass Transf.* **2003**, *46*, 4467–4481. [CrossRef]
18. Serageldin, A.A.; Sakata, Y.; Katsura, T.; Nagano, K. Thermo-hydraulic performance of the U-tube borehole heat exchanger with a novel oval cross-section: Numerical approach. *Energy Convers. Manag.* **2018**, *177*, 406–415. [CrossRef]
19. Jamshidi, N.; Mosaffa, A. Investigating the effects of geometric parameters on finned conical helical geothermal heat exchanger and its energy extraction capability. *Geothermics* **2018**, *76*, 177–189. [CrossRef]
20. Zhao, J.; Li, Y.; Wang, J. A review on heat transfer enhancement of borehole heat exchanger. *Energy Procedia* **2016**, *104*, 413–418. [CrossRef]
21. Acuña, J. Improvements of U-Pipe Borehole Heat Exchangers. Licenciate Thesis, KTH School of Industrial Engineering and Management, Stockholm, Sweden, 2010.
22. Ojala, M.; Ojala, M.; Ojala, K.; Ojala, H. Pipe Collector for Heat Pump Systems. U.S. Patent 9,546,802 (B2), 17 January 2017. Current Assignee: MUOVITECH AB, Brämhult, Sweden.
23. Bouhacina, B.; Saim, R.; Oztop, H.F. Numerical investigation of a novel tube design for the geothermal borehole heat exchanger. *Appl. Therm. Eng.* **2015**, *79*, 153–162. [CrossRef]
24. Bae, S.M.; Nam, Y.; Choi, J.M.; Lee, K.H.; Choi, J.S. Analysis on Thermal Performance of Ground Heat Exchanger According to Design Type Based on Thermal Response Test. *Energies* **2019**, *12*, 651. [CrossRef]
25. Fourier, J. *Theorie Analytique de la Chaleur, par M. Fourier*; Chez Firmin Didot, Père et Fils: Paris, France, 1822.
26. Ingersoll, L.R.; Zobel, O.J. *An Introduction to the Mathematical Theory of Heat Conduction*; Ginn and Company: Boston, MA, USA, 1913.
27. Chiasson, A.D. *Geothermal Heat Pump and Heat Engine Systems: Theory and Practice*; John Wiley & Sons Ltd.: Hoboken, NJ, USA, 2016.
28. Carslaw, H.S.; Jaeger, J.C. *Conduction of Heat in Solids*; Clarendon Press: Oxford, UK, 1959.
29. Eskilson, P. Thermal Analysis of Heat Extraction Boreholes. Ph.D. Thesis, University of Lund, Lund, Sweden, 1987.
30. Matthews, C.S.; Russell, D.G. *Pressure Buildup and Flow Tests in Wells*; Society of Petroleum Engineers of AIME: Dallas, TX, USA, 1967.
31. Lee, J. *Well Testing*; Society of Petroleum Engineers of AIME, SPE: Richardson, TX, USA, 1982.
32. Gehlin, S. Thermal Response test: Method Development and Evaluation. Ph.D. Thesis, Lulea University of Technology, Lulea, Sweden, 2002.
33. Claesson, J.; Eskilson, P. *Conductive Heat Extraction by a Deep Borehole*; Analytical Studies; Department of Mathematical Physics, University of Lund: Lund, Sweden, 1987.

34. Kurevija, T.; Macenić, M.; Borović, S. Impact of grout thermal conductivity on the long-term efficiency of the ground-source heat pump system. *Sustain. Cities Soc.* **2017**, *31*, 1–11. [CrossRef]
35. Šikić, K.; Basch, O.; Šimunić, A. *Basic geological map of SFRJ*; Croatian Geological Survey: Zagreb, Croatia, 1978.
36. Faculty of Mining, Geology and Petroleum Engineering, University of Zagreb, Study of water protection zones at water well Velika Gorica (original in Croatian: Elaborat Zaštitnih Zona Vodocrpilišta Velika Gorica). Available online: http://www.gorica.hr/dokumenti/elaborat_zastitnih_zona_velika_gorica.pdf (accessed on 20 March 2019).

Article

Long-Term Temperature Evaluation of a Ground-Coupled Heat Pump System Subject to Groundwater Flow

Nehed Jaziri [1], Jasmin Raymond [1,*], Nicoló Giordano [1] and John Molson [2]

1 Centre Eau terre Environnement, Institut National de la Recherche Scientifique, Québec, QC G1K 9A9, Canada; nehed.jaziri@yahoo.fr (N.J.); nicolo.Giordano@ete.inrs.ca (N.G.)

2 Département de Géologie et de Génie Géologique, Université Laval, Québec, QC G1V 0A6, Canada; john.molson@ggl.ulaval.ca

* Correspondence: jasmin.raymond@inrs.ca; Tel.: +1-418-654-2559

Received: 8 November 2019; Accepted: 16 December 2019; Published: 23 December 2019

Abstract: The performance of ground-coupled heat pump systems (GCHPs) operating under significant groundwater flow can be difficult to predict due to advective heat transfer in the subsurface. This is the case of the Carignan-Salières elementary school located on the south shore of the St. Lawrence River near Montréal, Canada. The building is heated and cooled with a GCHP system including 31 boreholes subject to varying groundwater flow conditions due to the proximity of an active quarry being irregularly dewatered. A study with the objective of predicting the borehole temperatures in order to anticipate potential operational problems was conducted, which provided an opportunity to evaluate the impact of groundwater flow. For this purpose, a numerical model was calibrated using a full-scale heat injection test and then run under different scenarios for a period of twenty years. The heat exchange capacity of the GCHP system is clearly enhanced by advection when the Darcy flux changes from 6×10^{-8} m s^{-1} (no dewatering) to 8×10^{-7} m s^{-1} (high dewatering). This study further suggests that even the lowest groundwater flow condition can be beneficial to avoid a progressive cooling of the subsurface due to the unbalanced building loads, which can have important impacts for design of new systems.

Keywords: geothermal; ground-coupled heat pump; numerical modelling; groundwater; FEFLOW; building; simulation

---

1. Introduction

Groundwater flow can have a significant impact on the long-term performance of vertical ground heat exchangers (GHEs), especially when the Darcy flux is greater than 1×10^{-7} m s^{-1} [1,2]. Flow in the subsurface can actually improve long-term GHE performance by dissipating heat injected into or extracted from the ground [3]. However, the design of ground-coupled heat pump (GCHP) systems is commonly based on the assumption of conductive heat transfer in the subsurface [4]. Although numerical tools are available to optimize the operation of a GCHP system under the influence of groundwater flow [5,6], an estimate of the specific groundwater flux affecting a GHE field can be difficult to define accurately. Alternatively, different authors have performed thermal response tests (TRTs) on a single GHE subject to groundwater flow to evaluate the equivalent subsurface thermal conductivity impacted by advection [7,8]. Practitioners commonly using such a heat conduction approach to simulate the long-term operating temperature of GHEs, based on an equivalent subsurface thermal conductivity assumption, have tried to cope with systems influenced by groundwater flow. This simplified approach can be useful, but neglects the fact that flow conditions can change over time.

The Carignan-Salières elementary school is an example of such a building, which is heated and cooled with a GCHP system operating under varying groundwater flow conditions because of its location within a kilometer of two quarries, one of which is being actively dewatered. In addition, because of the significant groundwater flow rates, upon installation the GHEs were backfilled with sand instead of using a geothermal grout. The grouting procedure was attempted but was not successful because the high flow rate most likely dispersed the fine grout particles into the fractures of the geological formations. This change of borehole filling material from the initial design plans, as well as the variable groundwater flow conditions, has likely affected the long-term operating temperature of the GHEs and their performance. The objective of this study is, therefore, to better understand the heat transfer mechanisms affecting the long-term operating temperature of the GHEs, including variable groundwater flow conditions, to anticipate potential operational problems.

Previous studies have been conducted to evaluate the impact of groundwater flow on the operating temperature of GHEs. For example, different authors simulated the operation of thermal energy storage systems with GHEs influenced by groundwater flow [9–11], while similar modelling was completed for GCHP systems by Dehkordi et al. [1]. However, most available studies are based on theoretical modelling exercises and have not been validated with operational data from systems that are influenced by significant groundwater flow conditions. This study provides a unique field case, where the GCHP system is located near an active quarry, at which groundwater flow conditions and ground thermal properties could be assessed. The GHE operating temperature could thus be accurately predicted based on a large-scale heat injection test carried out for the whole bore field, rather than a single GHE used for a TRT. Long-term numerical simulations of the system temperature under different field-based scenarios were carried out for a period of twenty years. A comparison between the simulated scenarios allowed a quantitative evaluation of the ground physical parameters affecting the GHE temperature, including the site hydraulic gradient. The knowledge gained can be used to improve the design and construction of new GCHP systems under the influence of groundwater flow.

2. Site Description

The Carignan-Salières elementary school is located on the south shore of the St. Lawrence River near Montreal (Figure 1) and was built above the Nicolet Formation. This sedimentary rock formation belongs to the Loraine Group, known to have a thermal conductivity on the order of 3 W m^{-1} K^{-1} [12–14], and is part of the St. Lawrence Lowlands sedimentary basin [15]. From a structural perspective, the St. Lawrence Lowlands are characterized by a series of normal faults extending from the southwest to the northeast, and which dip toward the southeast [16,17].

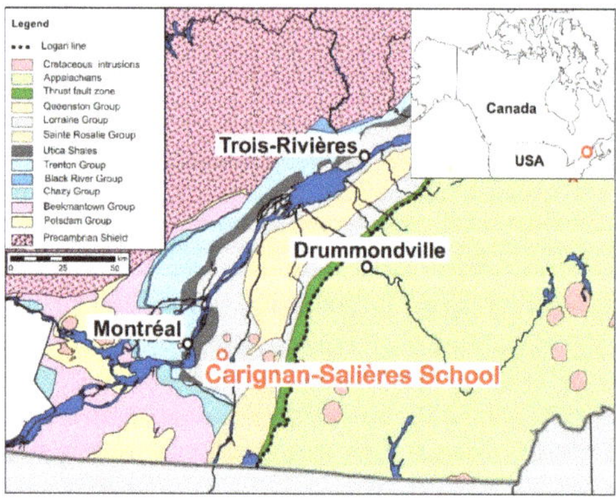

Figure 1. Simplified geological map of the area showing the location of the Carignan-Salières elementary school.

2.1. Geological and Hydrogeological Setting

The Nicolet Formation, hosting the school GHE field, consists of sequences of silty gray shale, with interbedded sandstone, siltstone and limestone [18]. Gabbro dykes, which are observed in the school area, are oriented EW and cut the stratigraphic sequence [19–21]. Most of the dykes are sub-vertical and have a varying thickness of 0.5 to more than 20 m [22]. One of the two quarries near the school is currently active and water is pumped irregularly (Figure 2). Regional aquifers in this area of the south shore of the St. Lawrence River are hosted in fractured rocks. The direction of groundwater flow at the site is oriented toward the active quarry where water is being pumped. The average hydraulic conductivity of the host rock reported in the area is 10^{-5} m s^{-1}, varying from 3.2×10^{-7} to 1.6×10^{-4} m s^{-1}, and the annual net recharge of the aquifer is approximately 100 mm y^{-1} [23]. The town of Carignan has a humid continental climate with an annual average air temperature of 5.9 °C and an amplitude of 30 °C [24]. The heating period is mostly from October to June while the cooling period is from July to September.

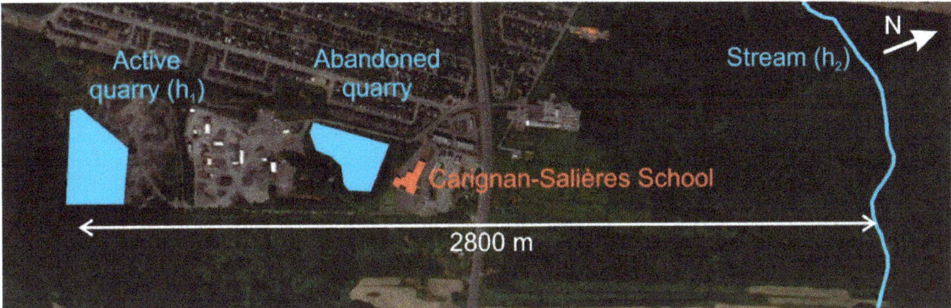

Figure 2. Satellite image of the Carignan-Salières school building and neighboring quarries.

2.2. Ground Heat Exchanger Characteristics

The GCHP system consists of thirty-one GHEs connected to fifty heat pumps distributed in the school building. Each heat pump has a net heating capacity of 3.62 to 44.2 kW. The system has sufficient capacity to supply the entire heating and cooling loads of the building. The GHEs are spaced by 6 m and each borehole is ~152 m deep. A high-density polyethylene single U-pipe with omega-shaped spacers constitutes the GHEs. During the installation, the boreholes could not be sealed with grout because groundwater flow along the intersecting fractures flushed the fine particles from the grout mixture. The boreholes were consequently filled with olivine sand instead of a common thermally-enhanced grout made of bentonite and sand. The olivine sand has a thermal conductivity of 1.75 W m^{-1} K^{-1}, a value that was measured in the laboratory during a previous study [25]. The heat carrier fluid is a mixture of water and propylene glycol at 25 vol. % and circulates in the closed loops with an average flow rate of 1017 m^3 d^{-1} within the entire GHE field. The GHEs are connected in parallel so the flow rate in each borehole is similar and around 3.8×10^{-4} m s^{-1}. Two TRTs were performed in two different boreholes, by injecting heat in order to evaluate the in-situ thermal conductivity of the subsurface. The first TRT was carried out before the GHE installation and revealed a bulk subsurface thermal conductivity of 2.58 W m^{-1} K^{-1} [26]. The second TRT was conducted after the GHE installation and indicated a lower value of 2.27 W m^{-1} K^{-1} [27]. The difference between the two tests is believed to be due to water pumping in the neighbouring and active quarry, inducing changes in the groundwater flow regime near the school.

3. Methodology

Taking advantage of the two quarries, field activities were undertaken to estimate the thermal conductivity of the subsurface and the groundwater flow conditions near the school site using representative values characteristics of a heterogeneous system. Twelve rock samples (6 shales, 1 calcarenite and 5 gabbros) were collected in the quarries to be analyzed in the laboratory. Water levels in the two quarry lakes were measured with a global positioning system to calculate the local hydraulic gradient. A heat injection test was further carried out for the entire GHE field with the system operating in cooling mode during the summer. Field-based observations allowed developing a representative numerical model of the GHE field, taking into account hourly building heating and cooling loads transferred to the ground.

3.1. Laboratory Measurement of Thermal Conductivity

The thermal conductivity assessments were made with two different methods: (1) a needle probe for hard rocks [12,28–30] and (2) the modified transient plane source (MTPS) method for friable rocks [30–32]. The KD2 Pro unit [33], part of a standardized method under the ASTM D5334 norm, was used for the needle probe analysis with gabbro samples, while the C-therm heating plate following the ASTM D7984 norm was used for the MTPS analysis with shales and calcarenites [34]. Before making the measurements, the samples were immersed for 24 h in distilled water to saturate the samples and obtain a value representative of subsurface conditions.

The samples analyzed with the needle probe were cut and a hole was drilled in each sample to insert a 6 cm long and 4 mm thick needle injecting heat at a rate of 6 W m^{-1}; the probe was covered with thermal grease to reduce contact resistance. After completing a measurement with a reference polyethylene sample, each rock sample was analyzed in a controlled-temperature room with at least ten measurements. An interval of 1 h was adopted for thermal equilibrium to be restored and a correction factor was taken into account to consider the reference sample measurements. With the MTPS method, a polished surface of the soft rock samples was placed on the heating plate to evaluate its thermal conductivity according to the transient increase of temperature. The heating plate has a disk shape and the electric signal used for the heat source serves as a proxy for temperature. The power level applied to the heat source is 90 V. Thermal conductivity of each sample was measured five times and an average was then calculated to obtain the final value.

3.2. In-Situ Heat Injection Test

The heat injection test to evaluate the thermal response of the entire GHE field was carried out during a hot summer period in July 2015. The test was completed by using the cooling system at its full capacity over 16.9 days, while the school windows had been opened to allow the outdoor heat to enter the building. The cooling system was then stopped, and the heat carrier fluid was kept circulating in the loop to monitor the thermal recovery during an additional 13.3 days. The flow rate of the heat carrier fluid and the temperature at the inlet and outlet of the entire GHE field were measured during the test by using the temperature sensors and flowmeters installed in the mechanical room of the GCHP system near the heat pump inlets. The temperature sensors and flowmeter have an accuracy of ±0.1 °C and ±1.5%, respectively. The instruments allowed taking measurements every 30 s. The heat injection rate was calculated from the temperature and flow rate measurements.

3.3. Building and Ground Load Evaluation

The school building loads were simulated using the program eQuest, a graphical interface for the DOE-2 program [35]. The simulations were used to establish a thermal energy budget of the building, which depends essentially on the building dimensions, the construction and insulation materials, the size and number of windows and doors, the operation schedule, as well as the internal and external temperature. Heat losses and gains of the indoor spaces were evaluated hourly to determine heating and cooling loads every hour over a full year. Simulated building loads, used as inputs in the numerical GHE model, were converted to ground loads according to:

$$P_{\text{ground}} = P_{\text{building}} \frac{COP_{\text{heating}}-1}{COP_{\text{heating}}} \tag{1}$$

and

$$P_{\text{ground}} = P_{\text{building}} \frac{COP_{\text{cooling}}+1}{COP_{\text{cooling}}} \tag{2}$$

where P_{ground} and P_{building} [W] are the loads for the ground and the building, respectively, and COP_{heating} and COPcooling [–] are the heat pump coefficients of performance in heating and cooling modes, respectively. An average and constant *COP* for all heat pumps of the building throughout the simulation time was assumed for simplification.

3.4. GCHP System Simulation

Numerical simulations to calibrate the model using the short-term heat injection test and to subsequently evaluate the long-term operating temperature of the GHEs over 20 years were carried out with the finite element program FEFLOW [36]. This program was used because it allows simulating transient heat transfer and groundwater flow in 3D porous media hosting GHEs embedded as 1D elements. A general iterative finite element strategy is used to solve the overall flow and heat transfer problem coupling the ground with the GHEs [37]. The numerical approach of Eskilson et al. [38], which represents GHEs with equivalent resistances, was selected instead of the more general numerical approach of Al-Khoury et al. [39,40]; both being available in FEFLOW. This choice was made because the Eskilson approach requires less computational effort and has been demonstrated to be accurate for long-term predictions [37].

3.4.1. Governing Equations

The global fluid flow and heat transfer problem is solved in the form of fluid, mass and thermal energy balances for the subsurface *s* and the groundwater *gw*. The flow equation is described by [36]:

$$S_s \frac{\partial h}{\partial t} + \nabla \cdot \mathbf{q} = Q + Q_{\text{EOB}} \tag{3}$$

where S_s [m^{-1}] is the specific storage coefficient, h [m] is the hydraulic head, Q [m^3 s^{-1}] is a source/sink term for flow and EOB refers to the Extended Oberbeck-Boussineq approximation. In Equation (3), $\mathbf{q}$ [m s^{-1}] is the Darcy flux in the porous medium and is expressed with Darcy's law:

$$\mathbf{q} = -\mathbf{K}(\nabla h) \tag{4}$$

where K [m s^{-1}] is the hydraulic conductivity tensor. The heat transport equation with conductive and advective terms is described by [41]:

$$(\rho c)_s \frac{\partial T}{\partial t} = \nabla \left(\boldsymbol{\lambda} \nabla T - (\rho c)_{gw}\, \mathbf{q}\, T \right) + H \tag{5}$$

where λ [W m^{-1} K^{-1}] is the thermal conductivity, ρc [J m^{-3} K^{-1}] is the volumetric heat capacity and H [W m^{-3}] represents in this case heat sources and sinks that can be, for example, tied to the GHEs.

The GHE equations used in the present model apply for single U-pipe heat exchangers, that, according to the thermal resistance and capacity model of Bauer et al. [42], consist of four components, namely an inlet-pipe denoted with subscript *il*, an outlet-pipe with subscript *ol* and two filling material zones (grout) with exponent or subscript *g*, and can be written as [36]:

$$\frac{\partial}{\partial t}\left(\rho^2 c^{\mathrm{r}} T_{\mathrm{il}}\right) + \nabla\cdot(\rho^{\mathrm{r}} c^{\mathrm{r}} \mathbf{u}\, T_{\mathrm{il}}) - \nabla\cdot(2^{\mathrm{r}}\cdot \nabla T_{\mathrm{il}}) \;=\; H_{\mathrm{il}} \tag{6}$$

$$\frac{\partial}{\partial t}(\rho^{\mathrm{r}} c^{\mathrm{r}} T_{\mathrm{ol}}) + \nabla\cdot(\rho^{\mathrm{r}} c^{\mathrm{r}} \mathbf{u} T_{\mathrm{ol}}) - \nabla\cdot(\Lambda^{\mathrm{r}}\cdot \nabla T_{\mathrm{ol}}) = H_{\mathrm{ol}} \tag{7}$$

$$\frac{\partial}{\partial t}\left(\varepsilon_g \rho^g c^g T_{g1}\right) - \nabla\cdot\left(\varepsilon_g \boldsymbol{\lambda}^g \nabla T_{g1}\right) = H_{g1} \tag{8}$$

$$\frac{\partial}{\partial t}\left(\varepsilon_g \rho^g c^g T_{g2}\right) - \nabla\cdot\left(\varepsilon_g \boldsymbol{\lambda}^g \nabla T_{g2}\right) = H_{g2} \tag{9}$$

where ρ [kg m^{-3}] and c [J kg^{-1} K^{-1}] are the density and the specific heat of the heat carrier fluid r and grout g, respectively, T [K] is the pipe temperature, H [W m^{-3}] is the thermal sink or source term, $\mathbf{u}$ [m s^{-1}] is the vector of the heat carrier fluid velocity, Λ^{r} [W m^{-2} K^{-1}] is the thermal hydrodynamic dispersion tensor for the heat carrier fluid, $\boldsymbol{\lambda}^g$ [W m^{-1} K^{-1}] is the thermal conductivity of the grout and ε_g [−] is the volume fraction of the grout or filling material. The relations are used to express the heat exchange between the borehole with 1 U-pipe and the subsurface. The effect of the 1 U-pipe components is lumped into effective heat transfer coefficients, which represent the sum of thermal resistances between the different components of the GHE elements [37].

3.4.2. Model Geometry and Properties

The surface area of the numerical GHE model is 500 m × 500 m and extends from ground surface to a depth of 300 m, divided into 6 layers of 50 m each. The same lithological characteristics of the subsurface, such as hydraulic conductivity, thermal conductivity of both fluid and solids, as well as porosity are assigned to the six layers that have a different initial temperature based on the geothermal gradient. The model mesh consists of 3D triangular prismatic elements for a total of 195,720 elements and 114,450 nodes. A 2D horizontal-plane mesh was built and then extended in 3D to cover the entire domain (Figure 3). Each GHE was surrounded by 6 nodes. Thirty-one GHEs were added to the domain, following the same layout as at the Carignan-Salières elementary school. In the plan view, the triangular mesh dimensions are up to approximately 4 m at the boundaries of the model (Figure 3), decreasing to 0.06 m around each GHE.

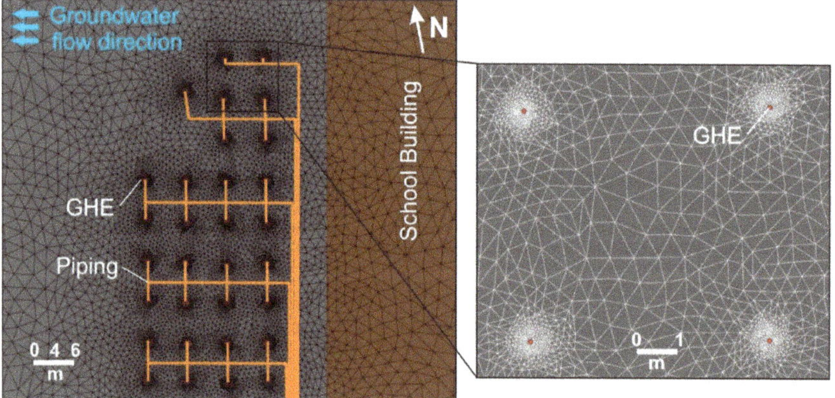

Figure 3. Plan view of the triangular prismatic mesh with details on the discretization around the ground heat exchangers (GHEs).

3.4.3. Initial and Boundary Conditions

Type-1 constant hydraulic heads with different values according to chosen simulation scenarios were imposed on the eastern and western boundaries of the model. This head gradient was applied to represent local groundwater flow conditions due to pumping in the active quarry, which is 1 km west (down-gradient) of the school location. The bottom surface was set impermeable and an annual net recharge of 100 mm y^{-1} was imposed at the top surface [23]. The initial ground temperature was assigned to each layer according to the geothermal gradient of the area (Figure 4), which is equal to 23.1 °C km^{-1} [12]. A fixed temperature boundary condition was used at the top of the model, while a constant heat flux was imposed at the bottom of the model in order to represent the Earth's heat flux in the St. Lawrence lowlands, which is assumed equal to 35 mW m^{-1} [43]. The lateral side boundaries were considered as Type-2 for heat transfer, with a zero temperature gradient assuming no conductive heat transfer. The GHEs in FEFLOW are Type-4 boundary conditions for which two approaches were used for the simulations. First, the inlet temperature was specified with an average flow rate of 1076 m^3 d^{-1} for the short-term calibration simulations to reproduce the outlet temperature during the heat injection test. Second, a total flow rate of 1017 m^3 d^{-1}, anticipated for the long-term operation of the system, and ground loads defined from the building simulation and the assumed heat pump *COPs* (Equations (1) and (2)), were assigned to predict the operating temperature of the GHE field for twenty years. For all numerical simulations, including model calibration, hourly time steps were used to ensure accurate results.

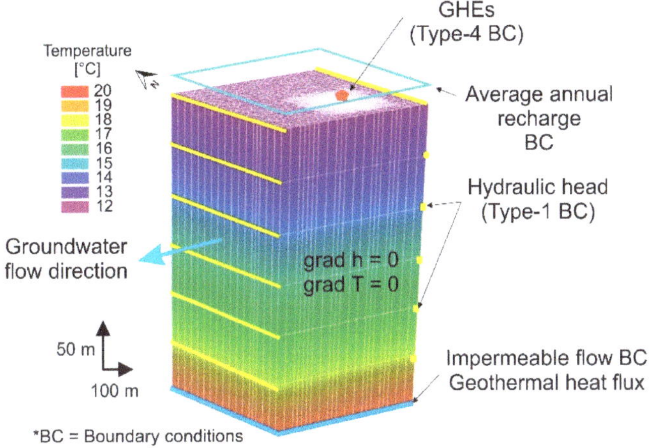

Figure 4. 3D model showing the boundary conditions and initial temperature for each layer.

3.4.4. Model Calibration

The model was calibrated to reproduce the outlet temperature recorded during the heat injection test. Parameters with the highest uncertainty, which are the thermal conductivity of the subsurface solids and the borehole backfill material, as well as the subsurface hydraulic conductivity, were manually adjusted to provide the best match between measured and simulated temperature. The objective was to identify possible ranges for uncertain parameters, which have an influence on the operating GHE temperature. All other parameters were kept constant during the calibration (Table 1), which are basically linked to the subsurface and the GHE entities. The constant parameters are the GHE pipe spacing, inlet and outlet pipe diameters, pipe thermal conductivity, pipe thickness, heat carrier fluid thermal conductivity and density, volumetric heat capacity of the fluid and subsurface solids, thermal conductivity of the fluid and specific storage of the subsurface.

Table 1. Constant parameters used for ground-coupled heat pump system (GCHP) simulations.

Parameter	Value
Subsurface	
Specific storage	10^{-4} m^{-1}
Volumetric heat capacity of water	4.2×10^6 J m^{-3} K^{-1}
Volumetric heat capacity of subsurface solids	2.52×10^6 J m^{-3} K^{-1}
Thermal conductivity of water	0.65 W m^{-1} K^{-1}
GHE	
Pipe spacing	0.10 m
Inlet and outlet pipe diameter	0.032 m
Pipe thermal conductivity	0.39 W m^{-1} K^{-1}
Pipe wall thickness	0.0038 m
Heat carrier fluid thermal conductivity	0.48 W m^{-1} K^{-1}
Heat carrier fluid density	1033 kg m^{-3}

4. Results

4.1. Groundwater Flow Conditions

Field observations combined with GHE drilling reports allowed to define a geological cross-section used as a conceptual model (Figure 5). The 2800 m cross-section is bounded by a stream and the active quarry that are considered constant-head hydraulic boundaries defining the water table at these locations. The thickness of the unconsolidated sediment cover was determined from well logs of the bore field [26]. The abandoned quarry closest to the school is 189 m down-gradient from the GHE field, while the active quarry is approximately 1 km down-gradient. The water level depth in the abandoned quarry was 21 m above sea level (asl) during the field work. Water is pumped irregularly in the active quarry to maintain the groundwater level below the excavation, which affects the local groundwater flow regime. The water level in the active quarry was measured at −3 m asl, but can vary by a few meters.

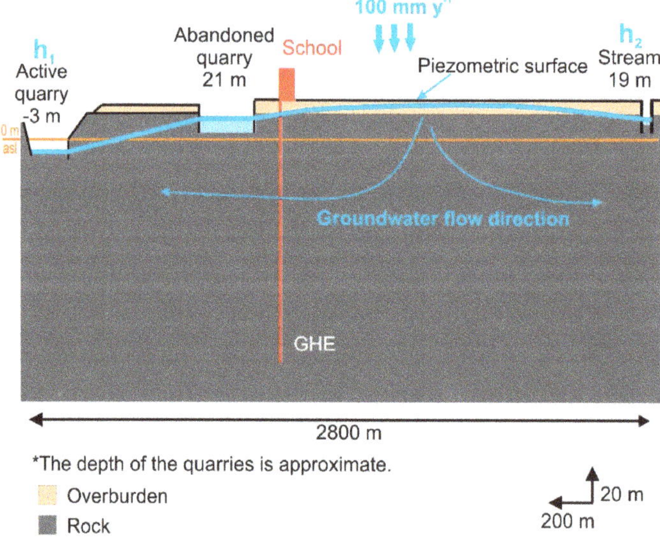

Figure 5. Conceptual model of the GHE field at the Carignan-Salières elementary school.

Steady-state groundwater flow in a simplified unconfined aquifer system with surface recharge was considered to calculate the level of the water table and the hydraulic gradient near the school site. A total distance of 2800 m separates the active quarry from the stream beyond the school, in the direction of the groundwater flow. Under an assumed constant hydraulic head maintained in the quarry and the stream, the hydraulic head near the GHE field was calculated using the Dupuit formulation according to [44]:

$$h^2 = h_2^2 - \frac{(h_2^2 - h_1^2)x}{L} + \frac{w}{K}(L - x)x \quad (10)$$

where h [m] is the hydraulic head between h_1 [m], the water level in the active quarry, and h_2 [m], the water level at the stream. The distance L [m] between the active quarry and the stream (2800 m), the average annual recharge w [m s^{-1}], the hydraulic conductivity of the subsurface K [m s^{-1}] and the distance x [m] measured from the stream (where x = 0) were considered to calculate the hydraulic head at the GHE field. Hydraulic conductivity and recharge were thus manually adjusted considering values reported by Carrier et al. (2013) to reproduce the hydraulic head measured in the abandoned quarry considered as an observation point. The conditions that best match the observation point (h = 20.7 m asl in the abandoned quarry) correspond to a hydraulic conductivity of 1.26 × 10^{-5} m s^{-1} and a net recharge of 100 mm y^{-1}, resulting in a hydraulic gradient at the GHE field of 0.008 m m^{-1}. Varying the hydraulic conductivity and the recharge rate with plausible literature values revealed a low sensitivity of the hydraulic gradient at the GHE field that changed by about 2.5 × 10^{-3} m m^{-1}.

4.2. Subsurface Thermal Conductivity

Gabbro samples taken from a dyke in the quarry and analyzed with both the needle probe and the MTPS have a thermal conductivity below 2.0 W m^{-1} K^{-1}, which is lower than the shales and calcarenites, due to the mineralogy of the intrusive rock containing abundant feldspars (Table 2). Thermal conductivity of gabbro was measured in another case study and ranged between 1.65 and 2.29 W m^{-1} K^{-1} [45], which is consistent with our results. Shale samples have an average thermal conductivity ranging between 2.4 and 2.9 W m^{-1} K^{-1} (Table 2), which is in agreement with a value of 2.8 W m^{-1} K^{-1} measured for other samples of similar lithology in the St. Lawrence lowlands [10,12,44,46]. The calcarenite sample has the highest thermal conductivity among all the samples collected in the quarry, with a value of 3.5 W m^{-1} K^{-1}. Results obtained from the laboratory analyses were considered to constrain the range of the solids thermal conductivity between 2 and 3 W m^{-1} K^{-1} in the subsequent numerical model since the thermal response tests reported in Section 2.2 were believed to be affected by groundwater flow.

Table 2. Average thermal conductivity of rock samples from the active quarry.

Rock Type	Measurement Method	Number of Samples Analyzed	Average Thermal Conductivity (W m^{-1} K^{-1})
Gabbro	Needle probe	3	1.87
Gabbro	MTPS	2	1.82
Calcarenite	MTPS	1	3.58
Dark shale	MTPS	3	2.42
Light shale	MTPS	3	2.85

4.3. In-Situ Heat Injection Test

The heat injection test was carried out at an average heat injection rate of 305 kW for the entire bore field or 9.8 kW per borehole (Figure 6). The total average flow rate was 39.9 L min^{-1}. The injection period lasted 406 h with GHE water temperature fluctuating between 15 °C to over 35 °C. Monitoring of the temperature recovery, where the fluid is kept circulating in the GHE field without heat injection, lasted 318 h and the GHE water temperature eventually recovered to about 13 °C.

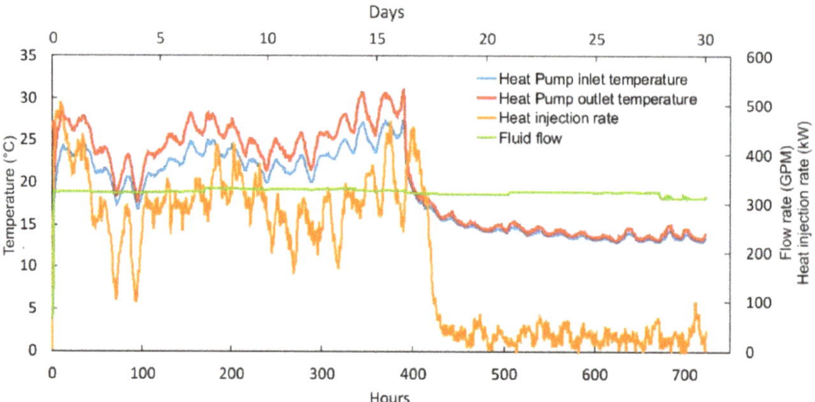

Figure 6. Fluid temperature recorded at the inlet and outlet of the GHE field during the heat injection test. Flow rate and heat injection rate are also presented.

4.4. GCHP System Simulation

4.4.1. Ground Loads

The building simulation using eQuest revealed that the total annual heating energy consumption of the school building is 171 MWh and the total cooling energy consumption is 119 MWh for a regular year of operation. The peak heating load of 494 kW occurs in January while the peak cooling load of 253 kW occurs during July. The building loads were converted to ground loads using Equations (1) and (2) assuming a *COP* of 4.7 and 4.1 in heating and cooling modes, respectively (Figure 7). The *COPs* were chosen according to the heat pump specifications provided by the manufacturer and represent average conditions. Heating loads are greater than cooling loads, which makes the ground loads unbalanced and can affect the long-term thermal response of the system.

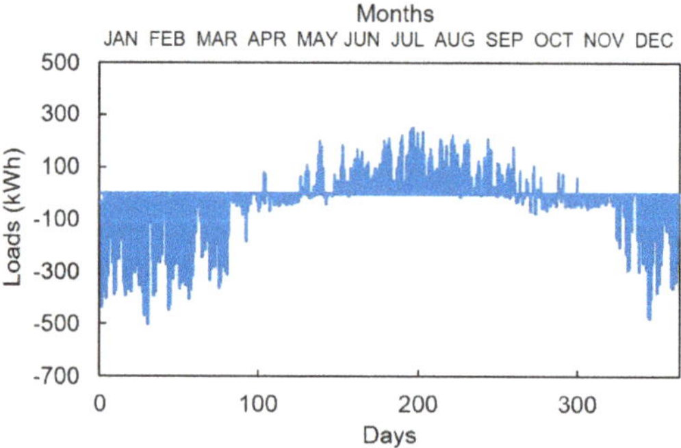

Figure 7. Simulated heating and cooling loads imposed on the GHE field of the school building calculated with eQuest from January to the end of December.

4.4.2. Model Calibration

The total duration for the calibration simulation was 724 h, which corresponds to the duration of the heat injection test, including the heat injection and the thermal recovery periods. Calibration parameters were adjusted one at a time (Table 3) until the calibrated model best reproduced the GHE outlet temperature with a maximum error of 2% (Figure 8). Calculation time using a workstation with a 3.3 GHz-core processor and 32Gb RAM was about 30 min for the calibration simulations.

Table 3. Range of parameter uncertainty in the calibration simulations.

Calibration Parameter	Possible Range	Chosen Value
K_x (m s^{-1})	10^{-5}–10^{-3}	10^{-4}
K_y (m s^{-1})	10^{-7}–10^{-3}	10^{-4}
K_z (m s^{-1})	10^{-9}–10^{-3}	10^{-6}
λ borehole filling material (W m^{-1} K^{-1})	1.5–1.9	1.75
λ host rock solids (W m^{-1} K^{-1})	2.1–2.4	2.4
Porosity (%)	3–5	3

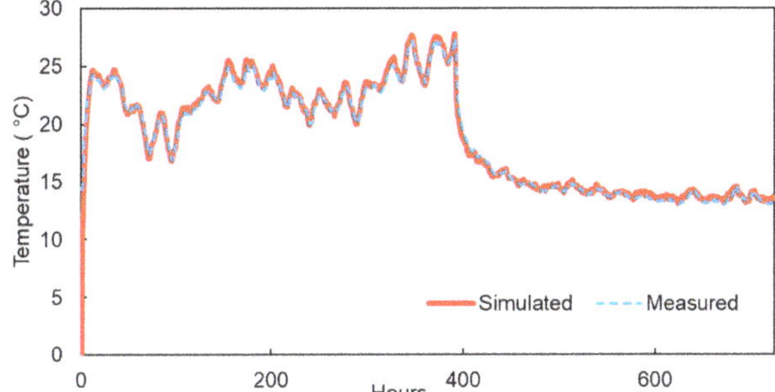

Figure 8. Best match between measured and simulated GHE outlet temperature during the heat injection test.

4.4.3. System Operation—Twenty-Year Simulations

The eight simulation scenarios used to evaluate the long-term operating GHE temperature at the Carignan-Salières elementary school were based on field observations and calibration results, varying key parameters one at a time to determine the impact of associated heat transfer mechanisms (Table 4). The start time of the simulation was the month of September, when the GHE system was put in operation. Each simulation took 5 days using the same workstation as reported above for the calibration simulations. Scenario A is a base case scenario, considering measured data for the thermal conductivity of the solids and the borehole filling material as well as hydraulic conductivity of the host rock reported in the literature [23] and a moderate hydraulic gradient, when compared to the hydraulic gradient measured during the field work. Scenario B was defined according to the best-fit parameters identified with the calibration and has a small change of solids thermal conductivity when compared to Scenario A. Only the thermal conductivity of the borehole filling material was changed in Scenarios C and D with respect to the base case. Scenarios E and F focused on thermal conductivity of the host rock solids, while Scenarios G and H were run to verify the influence of groundwater flow with a change of hydraulic head at the eastern and western boundaries of the model (Figure 4).

Table 4. Scenarios considered for GCHP system simulations at the Carignan-Salières elementary school.

Scenario	Hydraulic Head at Lateral Boundaries (m)	Hydraulic Gradient (m m^{-1})	Thermal Conductivity of Subsurface Solids (W m^{-1} K^{-1})	Thermal Conductivity of Borehole Filling Material (W m^{-1} K^{-1})
A	26–24	0.002	2.5	1.75
B	26–24	0.002	2.4	1.75
C	26–24	0.002	2.4	1.90
D	26–24	0.002	2.4	1.50
E	26–24	0.002	2.0	1.75
F	26–24	0.002	3.0	1.75
G	26–25.7	0.0006	2.4	1.75
H	26–22	0.008	2.4	1.75

Simulation results for Scenarios A and B do not show significant differences in the simulated GHE fluid temperature at both the outlet and inlet (Figure 9 and Table 5). The small decrease of the subsurface thermal conductivity in Scenario B did not have a significant impact on the results. Scenario D shows a GHE fluid temperature that is 0.4 °C higher than the maximum and lower than the minimum temperature when compared to Scenario C, which is due to the decrease of the thermal conductivity of the backfilling material. A low thermal conductivity of the subsurface (Scenario E) affects the maximum and the minimum GHE fluid temperature by up to 0.9 °C, when compared to the case with a high thermal conductivity (Scenario F).

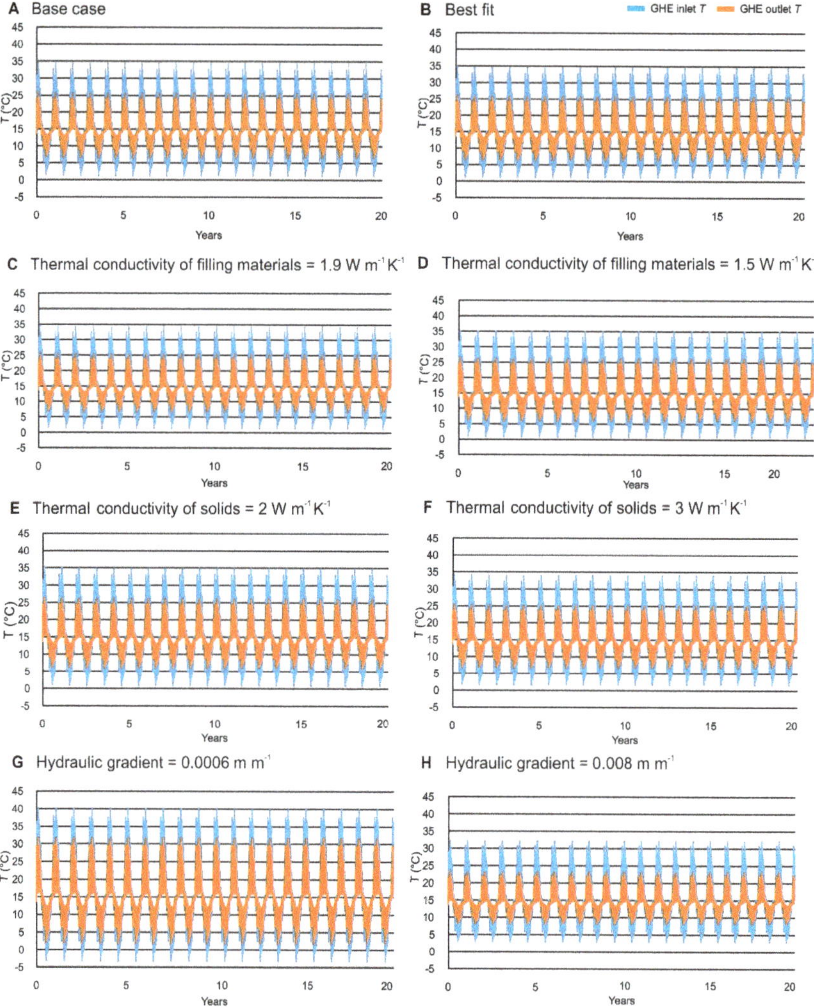

Figure 9. Simulated GHE inlet and outlet temperature for twenty years of operation at the Carignan-Salières school building according to different subsurface scenarios.

Table 5. Minimum and maximum GHE inlet temperature for the different scenarios during the 20th year of simulation. The variations given in percentage of the input parameters (Table 4) and the simulated temperature refer to scenario A.

Scenario	Hydraulic Gradient	Thermal Conductivity of Subsurface Solids	Thermal Conductivity of Borehole Filling Material	GHE Outlet Minimum/Maximum Temperature during Year 20			
				Heating Mode		Cooling Mode	
	(%)	(%)	(%)	(°C)	(%)	(°C)	(%)
A	–	–	–	6.4	–	25.6	–
B	0	−4	0	6.4	0	25.6	0
C	0	−4	9	6.5	2	25.4	−1
D	0	−4	−14	6.1	−5	26.1	2
E	0	−20	0	6.1	−5	26.0	2
F	0	20	0	6.9	8	25.1	−2
G	−70	−4	0	2.0	−69	31.7	24
H	300	−4	0	8.3	30	23.1	−10

Simulations with differences in hydraulic gradient imposed across the model domain show the most significant differences in GHE fluid temperature. In Scenario G, the GHE inlet temperature reaches 40 °C and drops to −3 °C and the GHE outlet temperature reaches more than 30 °C and drops to less than 2 °C. However, in Scenario H, the GHE inlet temperature reaches 32 °C and falls to only 3 °C while the GHE outlet temperature is between 23 °C and 8 °C. The maximum GHE fluid temperature in Scenario H is 8 °C lower than the maximum of Scenario G, while the minimum GHE fluid temperature is 6 °C higher than minimum of Scenario G. This trend can be explained by the higher imposed hydraulic gradient of 0.008, which was considered for Scenario H that represents conditions with significant pumping in the active quarry. This last scenario provides better heat exchange with the subsurface and therefore better GHE temperature or performance. The lower hydraulic gradient can represent a case where pumping in the active quarry is stopped or reduced, which has a negative impact on the GHE temperature and consequently a potential negative impact on the system performance. Overall, the thermal conductivity of the grouting material has a small impact on twenty-year predictions of the GHE inlet and outlet temperature. The thermal conductivity of the subsurface has a moderate impact on the maximum and minimum GHE temperature. The hydraulic gradient, which is thought to be affected by pumping water in the active quarry, has the greatest impact on heat exchange with the subsurface. Therefore, long-term GHE performance mostly depends on the local groundwater flow conditions. As the minimum GHE outlet temperature drops, the performance of the heat pump can decrease. Despite the significant differences between the simulated temperature of the heat carrier fluid in Scenarios G and H, the twenty-year simulations show an appropriate thermal response of the subsurface with constant temperature changes from year to year, although ground loads are unbalanced. A low groundwater flow rate appears enough to reduce the effect of unbalanced loads that can potentially cool the subsurface since this is not noticed in the long-term simulation. The main impact of groundwater flow is on yearly temperature changes. The minimum GHE outlet temperature dropped to 2 °C in Scenario G, which is still far from the minimal operating temperature of the heat pump system recommended by the manufacturer (−9.62 °C). However, the minimum GHE inlet temperature dropped to −3 °C. The freeze protection provided by the 25 vol. % propylene glycol solution circulating in the GHE is −10 °C and geothermal system designers recommend a minimum fluid temperature 5 to 7 °C higher than the freezing point of the solution. Therefore, under low groundwater flow conditions, care should be taken to follow the minimum operating system temperature during winter periods to avoid potential freezing problems at the GHE inlet.

5. Discussion

Previous studies that involved the numerical evaluation of GHE performance under the influence of groundwater flow identified the conditions where advection becomes the dominant heat transfer mechanism affecting GHE operation, which has been compared in terms of Darcy flux in the following discussion. For example, Chiasson et al. [47] concluded that groundwater flow can be expected to affect GHE performance, with finite element simulations of groundwater flow and heat transfer that indicated a higher minimum GHE water temperature when increasing Darcy flux from 1.9×10^{-6} to 1.9×10^{-5} m s^{-1}. In addition, Fujimoto et al. [48] studied a GCHP system where significant groundwater flow was shown to have more effect on the thermal response of GHEs when compared to the ground thermal conductivity, even under different building load scenarios. In less permeable subsurface layers, which had a groundwater Darcy flux of 2×10^{-10} m s^{-1}, Fujimoto et al. [48] showed that the operating temperature of a GHE exhibits more changes than that of the permeable layer having a Darcy flux of 6×10^{-7} m s^{-1}. Another study that used numerical simulations has shown that increasing the Darcy flux in a theoretical aquifer system from 10^{-9} to 10^{-7} m s^{-1} resulted in a longer and narrower thermal plume propagating from the GHE [1]. Further increasing the Darcy flux to 10−6 m s^{-1} made the thermal plume dramatically smaller. Dehkordi et al. [1] concluded that groundwater flow can substantially improve heat transfer by enhancing advection. This was confirmed by Ferguson [2], who provided a screening tool indicating that the boundary between GCHP systems with heat conduction versus advection as a dominant heat transfer mechanism is when the Darcy flux is on the order of 1×10^{-7} m s^{-1}.

The GCHP system at the Carignan-Salières elementary school has been operating near this boundary. Constant hydraulic heads imposed at the lateral boundaries of the model presented in this study were varied to evaluate the operating temperature of the GHE under specific Darcy fluxes varying from 8×10^{-7} to 6×10^{-8} m s^{-1}, which represented conditions with high to low groundwater flow rates in situations with and without pumping, respectively, in the neighbouring active quarry. The subsurface heat exchange capacity of the GCHP system appears to be enhanced by groundwater flow, as a consequence of pumping in the active quarry. The simulation carried out under the most likely operating conditions, assuming a thermal conductivity of the grouting material and the subsurface solids of 1.75 and 2.4 W m^{-1} K^{-1}, respectively, and with groundwater flow conditions that induce a Darcy flux of 2×10^{-7} m s^{-1} at the GHE field, indicates a reasonable operating GHE temperature for the next twenty years. However, if pumping activity ceases in the active quarry and the specific groundwater flux drops below 6×10^{-8} m s^{-1}, the minimum GHE outlet temperature can become critical and may drop closer to the minimum heat pump working temperature.

Underground thermal energy storage systems can also be affected by groundwater flow when the Darcy flux is on the same order of magnitude as that reported for GCHP systems (1×10^{-7} m s^{-1}) [11]. However, the impact of groundwater flow on underground thermal energy storage systems is negative because it increases heat loss around GHEs used as a heat storage medium. Giordano and Raymond [10] demonstrated that ~10% of the energy lost by the bore field of an underground energy storage system can be due to advection when the Darcy flux is 8×10^{-7} m s^{-1}, while this heat loss can be reduced by ~60 % if the connection and layout of the GHEs is designed considering the magnitude and direction of groundwater flow.

The present study further highlights the potential of groundwater flow to help operating GHEs with unbalanced ground loads, which were affected in this study by ~30 % more heat extraction due to greater heating than cooling of the building. The minimum GHE water temperature is expected to decrease year after year when simulating heating-dominated GCHP systems using a heat conduction approach to represent heat transfer in the subsurface. Jaziri et al. [49] previously simulated the operation of the GCHP system at the Carignan-Salières elementary school with such a heat conduction approach based on the HyGCHP program [50], where GHEs were represented with the duct storage model with no groundwater flow contribution [51]. The GHE minimum operating temperature at the beginning of the simulations was similar to simulations made with FEFLOW for the case with a low groundwater flow rate (Scenario G) but decreased by 4 to 6 °C over the twenty-years of system operation. On the other hand, GHE simulations made with FEFLOW and considering advection did not show a significant decrease of the minimum GHE water temperature over the expected life-span of the system, even with a low groundwater flow rate (Figure 9). This finding has important implications for GCHP system design. When calculating the required GHE length according to building energy needs, and considering the thermal state and properties of the subsurface [52], practitioners will ensure that the desired minimum GHE temperature is not reached before 10 to 20 years, assuming heat conduction only in the subsurface. This approach can overestimate the required GHE length if advection becomes significant. The present study suggests that this can be the case, even at a low groundwater flow rate with a Darcy flux on the order of 10–8 m s^{-1}. Possible approaches to design GCHP systems under the influence of groundwater flow are to assume heat conduction only, but calculate the GHE length for a one-year heat pulse [53] or consider advection with a moving line-source solution to complete the sizing calculation [54,55].

6. Conclusions

Field characterization and modelling work conducted at the Carignan-Salières elementary school has shown that the installed GCHP system can be influenced by the varying groundwater flow conditions due to dewatering of a neighbouring quarry located approximately 1 km down-gradient of the borehole field. The specific groundwater Darcy flux in the host rock, in which the GHE system is installed, has been estimated near 10^{-7} m s^{-1} at the time when the water levels in the nearby quarries were measured. The school building is constructed above fractured shale and limestone of the Nicolet Formation in the Lorraine Group of the St. Lawrence Lowlands and the groundwater flow is oriented westward toward the active quarry. Measurement of thermal conductivity of the rock samples collected in the active quarry revealed that the most abundant rock types have a thermal conductivity ranging from 1.8 to 3.6 W m^{-1} K^{-1}. A TRT conducted at the site, which reported an equivalent subsurface thermal conductivity of 2.58 W m^{-1} K^{-1} [26], was likely affected by groundwater flow. A heat injection test was therefore carried out with the purpose of evaluating the thermal response of the entire GHE field subject to groundwater flow. The data collected were used to calibrate a groundwater flow and heat transfer model built using FEFLOW to simulate the operating temperature of the GHEs. The numerical model was developed according to the characteristics of the subsurface and the GHE field to reproduce the heat injection test and then predict the long-term response according to different scenarios.

The parameters that impact the GHE operating temperature in order of increasing importance are: the thermal conductivity of the backfilling material, the thermal conductivity of the subsurface and the groundwater flow rate. The impact of the thermal conductivity of the borehole filling material and the subsurface (Scenarios B to F) on the inlet and outlet temperature of the GHEs was actually shown to be minor when compared to the possible change of groundwater flow conditions at the site (Scenarios G and H). The Carignan GCHP system is a unique field case with GHEs interfering with the groundwater drawdown around the quarry and where the local thermal and hydraulic conditions of the GCHP system have uncommonly been assessed at a large scale. The subsurface heat exchange capacity of the GCHP system is clearly enhanced by groundwater flow inducing significant advective heat transfer when the specific Darcy flux increases from 6×10^{-8} m s^{-1} (no dewatering) to 8×10^{-7} m s^{-1} (high dewatering). This study further suggests that even the lowest groundwater flow rates expected at the site can be beneficial to avoid a progressive cooling of the system over its expected lifetime due to the unbalanced heating and cooling loads.

Further research is necessary to define the conditions where groundwater flow can help to cope with unbalance ground loads. While it is clear that advection plays an important role on heat transfer when the Darcy flux is 10^{-7} m s^{-1} or higher, a lower groundwater flow rate may be sufficient to ensure a constant minimum GHE temperature year after year, depending on the magnitude of the unbalanced loads.

Author Contributions: N.J. did the field, laboratory and numerical modelling work and wrote the manuscript. J.R. and J.M. supervised the work and improved the original manuscript prepared by N.J., N.G. helped with numerical modelling and contributed to the revision of the manuscript. All authors have read and agreed to the published version of the manuscript.

Funding: This research was funded by a BMP-Innovation scholarship given to Nehed Jaziri from the Fonds de recherche du Québec—Nature et technologies, the Natural Sciences and Engineering Research Council (NSERC) and the company GBi as well as an NSERC Discovery grant given to Jasmin Raymond.

Acknowledgments: The support of GBi, particularly Maxime Boisclair and Jérôme Plante, and the Commission scolaire des Patriotes, through Jean-François Rondeau, who facilitated this study, are acknowledged.

Conflicts of Interest: The authors declare no conflict of interest.

Nomenclature

c	Specific heat capacity [J kg^{-1} K^{-1}]
COP	Coefficient of performance [–]
ε	Volume fraction [–]
H	Heat source/sink term [W m^{-3}]
h	Hydraulic head [m]
$\mathbf{K}$	Hydraulic conductivity tensor [m s^{-1}]
K	Hydraulic conductivity [m s^{-1}]
L	Distance [m]
P	Heating load [W]
Q	Hydraulic source/sink term [m^3 s^{-1}]
$\mathbf{q}$	Darcy flux tensor [m s^{-1}]
$\boldsymbol{\lambda}$	Thermal conductivity tensor [W m^{-1} K^{-1}]
λ	Thermal conductivity [W m^{-1} K^{-1}]
ρ	Density [kg m^{-3}]
ρc	Volumetric heat capacity [J m^{-3} K^{-1}]
S_s	Specific storage coefficient [m^{-1}]
T	Temperature [K] or [°C]
t	Time [s]
Λ	Thermal hydrodynamic dispersion tensor [W m^{-2} K^{-1}]
$\mathbf{u}$	Heat carrier fluid velocity tensor [m s^{-1}]
w	Annual recharge [m s^{-1}]

Subscripts

EOB	Extended Oberbeck-Boussineq approximation
g	Grout
gw	Groundwater flow
il	Inlet pipe
ol	Outlet pipe
r	Heat carrier fluid
s	Subsurface

Abbreviations

asl	Above sea level
E	East
GCHP	Ground-coupled heat pump
GHE	Ground heat exchanger
MTPS	Modified transient plane source
TRT	Thermal response test
W	West

References

1. Dehkordi, S.E.; Schincariol, R.A. Effect of thermal-hydrogeological and borehole heat exchanger properties on performance and impact of vertical closed-loop geothermal heat pump systems. *Hydrogeol. J.* **2014**, *22*, 189–203. [CrossRef]
2. Ferguson, G. Screening for heat transport by groundwater in closed geothermal systems. *Groundwater* **2015**, *53*, 503–506. [CrossRef] [PubMed]
3. Zanchini, E.; Lazzari, S.; Priarone, A. Long-term performance of large borehole heat exchanger fields with unbalanced seasonal loads and groundwater flow. *Energy* **2012**, *38*, 66–77. [CrossRef]
4. Bernier, M. Ground-coupled heat pump system simulation. *ASHRAE Trans.* **2001**, *107*, 605–616.
5. Fujii, H.; Itoi, R.; Fujii, J.; Uchida, Y. Optimizing the design of large-scale ground-coupled heat pump systems using groundwater and heat transport modeling. *Geothermics* **2005**, *34*, 347–364. [CrossRef]
6. Fujii, H.; Inatomi, T.; Itoi, R.; Uchida, Y. Development of suitability maps for ground-coupled heat pump systems using groundwater and heat transport models. *Geothermics* **2007**, *36*, 459–472. [CrossRef]
7. Signorelli, S.; Bassetti, S.; Pahud, D.; Kohl, T. Numerical evaluation of thermal response tests. *Geothermics* **2007**, *36*, 141–166. [CrossRef]
8. Bozdağ, Ş.; Turgut, B.; Paksoy, H.; Dikici, D.; Mazman, M.; Evliya, H. Ground water level influence on thermal response test in Adana, Turkey. *Int. J. Energy Res.* **2008**, *32*, 629–633. [CrossRef]
9. Bauer, D.; Heidemann, W.; Müller-Steinhagen, H.; Diersch, H.J.G. Modelling and simulation of groundwater influence on borehole thermal energy stores. Effstock. In Proceedings of the 11th International Conference on Energy Storage, Stockholm, Sweden, 14–17 June 2009.
10. Giordano, N.; Raymond, J. Alternative and sustainable heat production for drinking water needs in a subarctic climate (Nunavik, Canada): Borehole thermal energy storage to reduce fossil fuel dependency in off-grid communities. *Appl. Energy* **2019**, *252*, 113463. [CrossRef]
11. Nguyen, A.; Pasquier, P.; Marcotte, D. Borehole thermal energy storage systems under the influence of groundwater flow and time-varying surface temperature. *Geothermics* **2017**, *66*, 110–118. [CrossRef]
12. Bédard, K.; Comeau, F.-A.; Raymond, J.; Malo, M.; Nasr, M. Geothermal Characterization of the St. Lawrence Lowlands Sedimentary Basin, Québec, Canada. *Nat. Resour. Res.* **2017**, *27*, 479–502. [CrossRef]
13. Raymond, J.; Sirois, C.; Nasr, M.; Malo, M. Evaluating the geothermal heat pump potential from a thermostratigraphic assessment of rock samples in the St. Lawrence Lowlands, Canada. *Environ. Earth Sci.* **2017**, *76*, 83. [CrossRef]
14. Raymond, J.; BÉdard, K.; Comeau, F.-A.; Gloaguen, E.; Comeau, G.; Millet, E.; Foy, S. A workflow for bedrock thermal conductivity map to help designing geothermal heat pump systems in the St. Lawrence Lowlands, Québec, Canada. *Sci. Technol. Built Environ.* **2019**, *25*, 963–979. [CrossRef]
15. Brisebois, D.; Brun, J. *La Plate-Forme du Saint-Laurent et les Appalaches. Géologie du Québec; Gouvernement du Québec, Public Report MM 94 -01*; Ministère des ressources naturelles: Quebec City, QC, Canada, 1994; pp. 95–120.
16. Séjourné, S.; Dietrich, J.; Malo, M. Seismic characterization of the structural front of southern Quebec Appalachians. *Bull. Can. Pet. Geol.* **2003**, *51*, 29–44. [CrossRef]
17. Castonguay, S.; Lavoie, D.; Dietrich, J.; Laliberté, J.-Y. Structure and petroleum plays of the St. Lawrence Platform and Appalachians in southern Quebec: Insights from interpretation of MRNQ seismic reflection data. *Bull. Can. Pet. Geol.* **2010**, *58*, 219–234. [CrossRef]
18. Globensky, Y. *Géologie des Basses-Terres du Saint-Laurent*; Ministère de l'Énergie et des Ressources du Québec: Quebec City, QC, Canada, 1987.
19. Foster, J.; Symons, D.T.A. Defining a paleomagnetic polarity pattern in the Monteregian intrusives. *Can. J. Earth Sci.* **1979**, *16*, 1716–1725. [CrossRef]
20. Foland, K.A.; Gilbert, L.A.; Sebring, C.A.; Jiang-Feng, C. 40Ar/39Ar ages for plutons of the Monteregian Hills, Quebec: Evidence for a single episode of Cretaceous magmatism. *GSA Bull.* **1986**, *97*, 966–974. [CrossRef]
21. Feininger, T.; Goodacre, A.K. The eight classical Monteregian hills at depth and the mechanism of their intrusion. *Can. J. Earth Sci.* **1995**, *32*, 1350–1364. [CrossRef]
22. Hodgson, C.J. Monteregian Dike Rocks. Ph.D. Thesis, McGill University, Montreal, QC, Canada, 1969.

23. Carrier, M.-A.; Lefebvre, R.; Rivard, C.; Parent, M.; Ballard, J.-M.; Benoit, N.; Vigneault, H.; Beaudry, C.; Malet, X.; Laurencelle, M.; et al. *Portrait des Ressources en eau Souterraine en Montérégie Est, Québec, Canada*; PACES Final Report R-1433; Institut national de la recherche scientifique—Centre Eau Terre Environnement: Quebec City, QC, Canada, 2013.
24. Environment Canada Climate Normals 1981–2010. Available online: http://www.climat.meteo.ec.gc.ca/climate_normals/index_e.html (accessed on 20 November 2019).
25. Côté, J.; Masoumifard, V.; Noreau, É. Optimizing the thermal conductivity of alternative geothermal grouts. In Proceedings of the 65th Canadian Geotechnical Conference, Winnipeg, MB, Canada, 30 September–3 October 2012.
26. Leblanc, Y. *Analyse d'un essai de réponse thermique—Ville de Carignan*; Internal Report; Richelieu Hydrogéologie: Richelieu, QC, Canada, 2012.
27. Renaud, B. *Évaluation de la conductivité du sol—École Carignan-Salières*; Internal Report; Géo-énergie: Boucherville, QC, Canada, 2013.
28. Bristow, K.L.; White, R.D. Comparison of single and dual probes for measuring soil thermal properties with transient heating. *Aust. J. Soil Res.* **1994**, *32*, 447–464. [CrossRef]
29. Barry-Macaulay, D.; Bouazza, A.; Singh, R.M.; Wang, B.; Ranjith, P.G. Thermal conductivity of soils and rocks from the Melbourne (Australia) region. *Eng. Geol.* **2013**, *164*, 131–138. [CrossRef]
30. Raymond, J. Colloquium 2016: Assessment of subsurface thermal conductivity for geothermal applications. *Can. Geotech. J.* **2018**, *55*, 1209–1229. [CrossRef]
31. Di Sipio, E.; Galgaro, A.; Destro, E.; Teza, G.; Chiesa, S.; Giaretta, A.; Manzella, A. Subsurface thermal conductivity assessment in Calabria (southern Italy): A regional case study. *Environ. Earth Sci.* **2014**, *72*, 1383–1401. [CrossRef]
32. Harris, A.; Kazachenko, S.; Bateman, R.; Nickerson, J.; Emanuel, M. Measuring the thermal conductivity of heat transfer fluids via the modified transient plane source (MTPS). *J. Therm. Anal. Calorim.* **2014**, *116*, 1309–1314. [CrossRef]
33. *Decagon Services KD2 Pro thermal properties analyzer*; Instruction manual, Decagon Services: Pullman, WA, USA, 2016.
34. *C-Therm Technologies TCi Thermal Conductivity Analyzer*; Instruction Manual; C-Therm Technologies: Fredericton, NB, Canada, 2016.
35. Hirsch, J.J. *DOE-2.2 Building Energy Use and Cost Analysis Program. Volume 1: Basics*; Lawrence Berkely National Laboratory: Berkeley, CA, USA, 2004; Available online: http://doe2.com/DOE2/index.html (accessed on 20 November 2019).
36. Diersch, H.J.G.; Bauer, D.; Heidemann, W.; Schatzl, P. *FEFLOW White Papers Vol. 5*; DHI-WASY GmbH: Berlin, Germany, 2010.
37. Diersch, H.-J.G.; Bauer, D.; Heidemann, W.; Rühaak, W.; Schätzl, P. Finite element modeling of borehole heat exchanger systems: Part 2. Numerical simulation. *Comput. Geosci.* **2011**, *37*, 1136–1147. [CrossRef]
38. Eskilson, P.; Claesson, J. Simulation model for thermally interacting heat extraction boreholes. *Numer. Heat Transf.* **1988**, *13*, 149–165. [CrossRef]
39. Al-Khoury, R.; Bonnier, P.G.; Brinkgreve, R.B.J. Efficient finite element formulation for geothermal heating systems. Part I: Steady state. *Int. J. Numer. Methods Eng.* **2005**, *63*, 988–1013. [CrossRef]
40. Al-Khoury, R.; Bonnier, P.G. Efficient finite element formulation for geothermal heating systems. Part II: Transient. *Int. J. Numer. Methods Eng.* **2006**, *67*, 725–745. [CrossRef]
41. Rühaak, W.; Renz, A.; Schätzl, P.; Diersch, H.J.G. *Numerical Modelling of Geothermal Applications*; World Geothermal Congress: Bali, Indonesia, 2010.
42. Bauer, D.; Heidemann, W.; Müller-Steinhagen, H.; Diersch, H.-J.G. Thermal resistance and capacity models for borehole heat exchangers. *Int. J. Energy Res.* **2011**, *35*, 312–320. [CrossRef]
43. Saull, V.A.; Clark, T.H.; Doig, R.P.; Butler, R.B. Terrestrial Heat Flow in the St. Lawrence Lowland of Québec. *Can. Min. Metall. Bull.* **1962**, *65*, 63–66.
44. Fetter, C.W. *Applied Hydrogeology*, 4th ed.; Prentice Hall: Upper Saddle River, NJ, USA, 2001.
45. Pasquale, V.; Verdoya, M.; Chiozzi, P. Measurements of rock thermal conductivity with a Transient Divided Bar. *Geothermics* **2015**, *53*, 183–189. [CrossRef]
46. Nasr, M.; Raymond, J.; Malo, M.; Gloaguen, E. Geothermal potential of the St. Lawrence Lowlands sedimentary basin from well log analysis. *Geothermics* **2018**, *75*, 68–80. [CrossRef]

47. Chiasson, A.; Rees, S.J.; Spitler, J.D. A preliminary assessment of the effects of groundwater flow on closed-loop ground-source heat pump systems. *ASHRAE Trans.* **2000**, *106*, 380–393.
48. Fujimoto, M.; Fujii, H.; Nagano, K.; Hiromatsu, A. Numerical modelling of large-scale cluster of vertical ground heat exchangers. *GRC Trans.* **2011**, *35*, 1101–1105.
49. Jaziri, N.; Raymond, J.; Boisclair, M. Performance evaluation of a ground coupled heat pump system with a heat injection test analysis. In Proceedings of the 69th Canadian Geotechnical Conference, Vancouver, BC, Canada, 2–5 October 2016.
50. Hackel, S.; Pertzborn, A. Building on experience with hybrid ground source heat pump systems. *ASHRAE Trans.* **2011**, *117*, 271–278.
51. Hellström, G. Ground Heat Storage: Thermal Analysis of Duct Storage Systems. Ph.D. Thesis, Department of mathematical physics, University of Lund, Lund, Sweden, 1991.
52. Philippe, M.; Bernier, M.; Marchio, D. Sizing calculation spreadsheet–Vertical geothermal borefields. *ASHRAE J.* **2010**, *52*, 20–28.
53. Morrison, A.D. GS2000tm Software. In Proceedings of the Fourth International Conference on Heat Pumps in Cold Climates, Alymer, QC, Canada, 17–18 August 2000.
54. Chiasson, A.; O'Connell, A. New analytical solution for sizing vertical borehole ground heat exchangers in environments with significant groundwater flow: Parameter estimation from thermal response test data. *HVAC R Res.* **2011**, *17*, 1000–1011.
55. Molina-Giraldo, N.; Blum, P.; Zhu, K.; Bayer, P.; Fang, Z. A moving finite line source model to simulate borehole heat exchangers with groundwater advection. *Int. J. Therm. Sci.* **2011**, *50*, 2506–2513. [CrossRef]

Article

Energy Efficiency of a Heat Pump System: Case Study in Two Pig Houses

Hannah Licharz [1,2,*], Peter Rösmann [3], Manuel S. Krommweh [1], Ehab Mostafa [4] and Wolfgang Büscher [1]

1 Institute of Agricultural Engineering, University of Bonn, Nußallee 5, 53115 Bonn, Germany; krommweh@uni-bonn.de (M.S.K.); buescher@uni-bonn.de (W.B.)
2 Institute of Animal Science, University of Bonn, Endenicher Allee 15, 53115 Bonn, Germany
3 AGRAVIS Futtermittel GmbH (Animal Feeding Company), Industrieweg 110, 48115 Münster, Germany; Peter.Roesmann@agravis.de
4 Agricultural Engineering Department, Faculty of Agriculture, Cairo University, El-Gammaa Street, Giza 12613, Egypt; ehababdelmoniem@hotmail.com
* Correspondence: hlicharz@uni-bonn.de; Tel.: +49-228-73-3081

Received: 19 December 2019; Accepted: 23 January 2020; Published: 4 February 2020

Abstract: This study describes a 70-day investigation of three identical groundwater heat pumps (GWHP) for heating two pig houses located on the same farm in West Germany. Two of the three GWHPs were installed in a piglet-rearing barn, the third in a farrowing barn. All three heat pumps were fed from the same extraction well. The aim of this study was firstly the empirical performance measurement of the GWHP systems and secondly the energetic evaluation of the systems on barn level by calculating the coefficient of performance (COP). Three different assessment limits were considered in order to better identify factors influencing the COP. In total, the heat pumps supplied thermal energy of 47,160 kWh (piglet-rearing barn) and 36,500 kWh (farrowing barn). Depending on the assessment limit considered, the COP in piglet-rearing barn and farrowing barn ranged between 2.6–3.4 and 2.5–3.0, respectively. A significant factor influencing the COP is the amount of electrical current required to operate the groundwater feeding pump. The average groundwater flow rate was 168.4 m^3 d^{-1} (piglet-rearing barn) and 99.1 m^3 d^{-1} (farrowing barn). In conclusion, by using energy from groundwater, GWHPs have the potential to substitute fossil fuels, thus saving them and avoiding CO_2 emissions.

Keywords: renewable energy; groundwater; livestock building; heating; coefficient of performance; climate change; resources; sustainability

1. Introduction

Against the background of limited fossil energy resources, the efficient use of energy and the sustainable use of energy resources play an important role worldwide. In addition, the reduction of greenhouse gas emissions is being sought in the context of climate change. In 2018, total greenhouse gas emissions in Germany amounted to 865.6 million tons of CO_2 equivalents [1]. A total of 63.6 million tons of these emissions are attributable to agriculture [1]. This means that German agriculture was able to achieve a reduction of 4.1% compared with the previous year 2017 (66.3 million tons of CO_2 equivalents from agriculture) [1]. However, there is still a need for action in agriculture and all other sectors in order to achieve national and international climate targets.

The demand for fossil fuels also plays a major role. The higher demand of the fast-growing economies of developing countries has led to high prices and decreasing fossil energy resource [2]. Consequently, the use of renewable energies, such as biogas, wind, and photovoltaic energy [2], and the use of geothermal energy is an important alternative to fossil fuels or energy from fossil fuels in general, but also in agriculture in particular (cf. [3]). This is an important reason for using heat pumps by substituting fossil fuels with other energy sources. Another reason is to become more independent of the energy price.

In pig housing, a good indoor climate must be available to the animals to satisfy the legislation regarding animal welfare demands and the thermal and physical requirements to ensure building protection. Due to the high temperature demand of piglets, the energy consumption during the rearing phase is relatively high. Thus, the average yearly energy consumption under German conditions is approximately 300 kWh per sow, including piglet rearing [4]. In Germany, pigs are typically housed in insulated buildings with forced ventilation. To maintain thermal welfare, just-born piglets require 35–37 °C, 28–32 °C in the first week postpartum (pp) and 25–27 °C in the later rearing phase [5,6]. The optimal surface temperature in the lying area should be 38–39 °C in the first week pp; for older piglets, the floor temperature in the lying area should be kept at 33–36 °C [7]. When planning a pig house in Germany, regulation DIN 18910 [8] must be considered; this regulation describes the calculation methods used to estimate the heat demand at the house level considering the air volume flow under summer and winter conditions. An example calculation is described by Büscher et al. [9]. In livestock buildings, the energy losses by ventilation are relatively high, making up 70 to 90% of the building's total heat losses [10–12]. Heat recovery systems are used to reduce these energy losses through ventilation of mechanically ventilated livestock buildings to transfer heat of the exhaust air to the incoming fresh air [13] (cf. [10,14,15]). In addition, systems using renewable energy for heating are and have been investigated and evaluated under practical conditions. Different technologies for heating pig housings using renewable energy are available, for example, a modular housing system with an integrated geothermal heat exchanger [3] or heat pump.

Basically, heat pumps extract heat from an energy source, which can be air, groundwater, or soil [16] (cf. Figure 1) by evaporating a refrigerant. With this system, the refrigerant is contained in a closed circuit and is guided to a compressor. Due to the compressing process, the gas changes to the liquid phase, and the energy can be transferred to another medium (e.g., to water circulating in a floor-heating system). The working medium is refrigerant that evaporates at a relatively low temperature. It recirculates between the energy source and the heat pump, changing its aggregation condition by the processes of evaporating, compressing, liquefying, and expanding, as shown in Figure 1. This heat transfer from "cold" to "hot" does not run automatically. The system consumes electricity for circulation pumps and compressors [17,18].

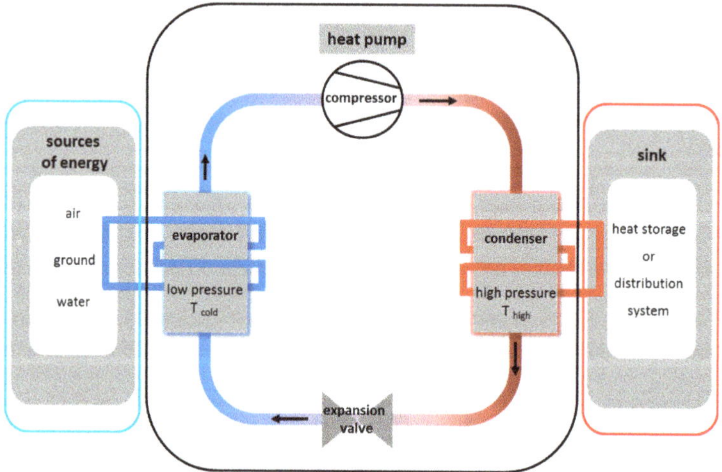

Figure 1. Principle and functional elements of a heat pump.

The temperature level of the heating system should be as low as possible (e.g., 35 °C) because of the decreasing efficiency of heat pumps for higher outgoing temperatures [18] (cf. [19]). Heat pumps can also be used to produce higher water temperatures in the heating systems (e.g., 55–65 °C) but only with low efficiency [18]. To make a comparison of different heat pumps regarding the COP (coefficient of performance) (see Equation (1)) on a test bench, the temperatures of the source and the sink medium have to be examined under defined conditions [20]. To determine the COP of heat pumps for planning aspects, the specific practical conditions of the installation must be regarded, considering other electricity consumers, such as the pumps for the circulation of hot water to distribute the heat energy inside the barn.

$$COP_{HP} = \frac{\dot{Q}_{HP}}{P_{HP}} \tag{1}$$

with COP_{HP}: coefficient of performance of heat pump; $\dot{Q}_{HP}$: heating performance of heat pump in kW; P_{HP}: electrical power consumption of heat pump in kW.

In heat pump systems, a distinction is made between open and closed systems. Open systems use groundwater or water from surface waters (pond, lake, or similar) as heat transfer medium, whereby the water is transported directly to the heat pump. After passing through the heat pump, the water is pumped into a rejection well or back into the surface waters. Closed systems are systems with a closed pipe system containing a heat transfer medium. The heat transfer medium circulates continuously through the pipe system and thus transports heat energy from the ground to the heat pump. Furthermore, heat pump systems differ in the design of the pipes, which can be arranged horizontally (laid close to the surface in the ground, 1–2 m deep) or vertically (reaching deep into the ground, up to 450 m deep [19,21]). In any case, a number of factors should be considered during the planning phase, which Omer summarizes well as follows: "geology and hydrogeology of the underground (sufficient permeability is a must for open systems), area and utilisation on the surface (horizontal closed systems require a certain area), existence of potential heat sources like mines, and the heating and cooling characteristics of the building(s)." [21] (p. 356). A detailed description of the various heat pump systems with their advantages and disadvantages is given in particular by Omar [21] and Ahmadi et al. [22].

When planning the use of a groundwater heat pump in Germany, it must be remembered that the use of groundwater is regulated by various laws. One important law is the Water Resources Act. Depending on the site, the Federal Mining Act may also apply. Every deep drilling requires the water law approval of the local district office (especially the water authority). It is checked whether the planned drilling site is located in a protected area (nature reserve, water protection area, or similar), as drilling is usually prohibited there. Other aspects that are normally checked are the quantity of groundwater available at the site as well as quality and flow velocity of the groundwater (cf. [17]). It may also be necessary to prepare a geological survey to clarify the soil conditions in deeper layers.

In Germany, but also in other European countries and worldwide, heat pumps of various designs are frequently used in residential buildings, office buildings, and in industry [22–25]. In agriculture, heat pumps are used, for example, to dry medicinal plants [26] or to heat greenhouses [27,28]. Not too many publications deal with the use of heat pumps for heating animal stables [17,29–36]. While Manolakos et al. [32] developed a model for a system, which uses heat pumps in a HVAC (heating, ventilation, and air conditioning) system for precise environment control in broiler houses (cf. [37]), only five of these publications [29,31,33,34,36] report on practical experiments with heat pump systems. The present study should help to close this gap.

Cremer [29] examined a farrowing barn with 40 animal places, in which a brine heat pump (15 kW nominal heating capacity; area collector) was used to heat the piglet nests. The heat pump produced 32,000 kWh thermal energy and had a COP of 4.0 in the one-year period of the study. The author states that the use of the heat pump saved 52% of energy costs compared to the use of an oil heating system. However, in calculating the COP, the author only considered the electricity consumption of the heat pump itself, but not that of the brine pump and the circulation pumps of the distribution system in the house. Furthermore, Cremer investigated an outdoor climate piglet-rearing house (1100 animal places) with a brine heat pump (16 kW nominal heating capacity; area collector) connected to a floor heating system in the cubicles [29]. During the 21-week study period, a total of 25,467 kWh of thermal energy was provided. Taking into account all relevant electricity consumers, 9961 kWh of electric power was required, resulting in a COP of 2.6 [29]. The potential of heat pump technologies in animal houses is anticipated to be high because temperature of the buffer reservoir of the floor-heating systems are relatively low (45 °C), and water-distribution systems are established in most cases [29]. Nevertheless, the requirements for a heat pump system in an animal house are different from those for residential or office buildings. In a pigsty, significantly higher temperatures in the animal area are to be achieved; for piglets between 25 °C and 37 °C (see paragraph above). Furthermore, mechanically ventilated livestock buildings are ventilated much more intensively than residential or office buildings, which is associated with higher heat losses via the heated exhaust air. Since a heat pump, depending on its dimensions, is not always able to cover a high heat requirement in the barn with cold outside air temperatures at the same time, it is not unusual that in practice another heating technology (hot air blower) is also used in order to still be able to guarantee the required barn interior temperature at these times.

The aim of this study was (1) the 70-day study of a farrowing barn and a piglet-rearing barn with identical heat pump types at the same location and (2) the energetic evaluation of the systems at barn level by calculating the coefficient of performance (COP). Three different assessment limits were considered in order to better identify factors influencing the performance. These are

1. COP_T: the thermal output was set in relation to the total amount of electricity used for pumping medium (ground water pump), heat pump itself (compressors), circulation pumps of the heat pump, and distribution system (circulation pumps for hot water distribution in the house).
2. COP_P: the thermal output was set in relation to the amount of electricity used for pumping medium (groundwater pump) and the heat pump itself (compressors).
3. COP_Z: the thermal output was set in relation to the amount of electricity used for the heat pump itself (compressors), circulation pumps of heat pump, and distribution system (circulation pumps for hot water distribution in the barn).

2. Materials and Methods

A case study from 8 December 2011 to 15 February 2012 was conducted to examine the studied heat pumps. The field experiments were performed in two forced-ventilated pig buildings (Figure 2) located in a district of Recklinghausen, West Germany. The district of Recklinghausen is located 440 km southwest of Berlin, 520 km northwest of Munich, 90 km north of Cologne, and 80 km east of Nijmegen (The Netherlands). These buildings were thermally insulated; the transition transfer coefficient of the entire building was 0.3 W m^{-2} K^{-1} (U-value) on average. According to the seasonal effects, three identical groundwater heat pumps (GWHPs) (Fighter 1330, NIBE Systemtechnik GmbH, Celle, Germany) were investigated. The first heat pump was installed in a farrowing barn with a total floor area of 3256 m^2 (88 m length × 37 m width), whereas the other two pumps were installed in the piglet-rearing barn with a total floor area of 2135 m^2 (61 m length × 35 m width). Each heat pump consisted of two modules (A and B). Both barns are a part of a farm with 740 sows, which are used for gilt reproduction. The farrowing and piglet-rearing barns were rebuilt away from the urbanized area in 2009.

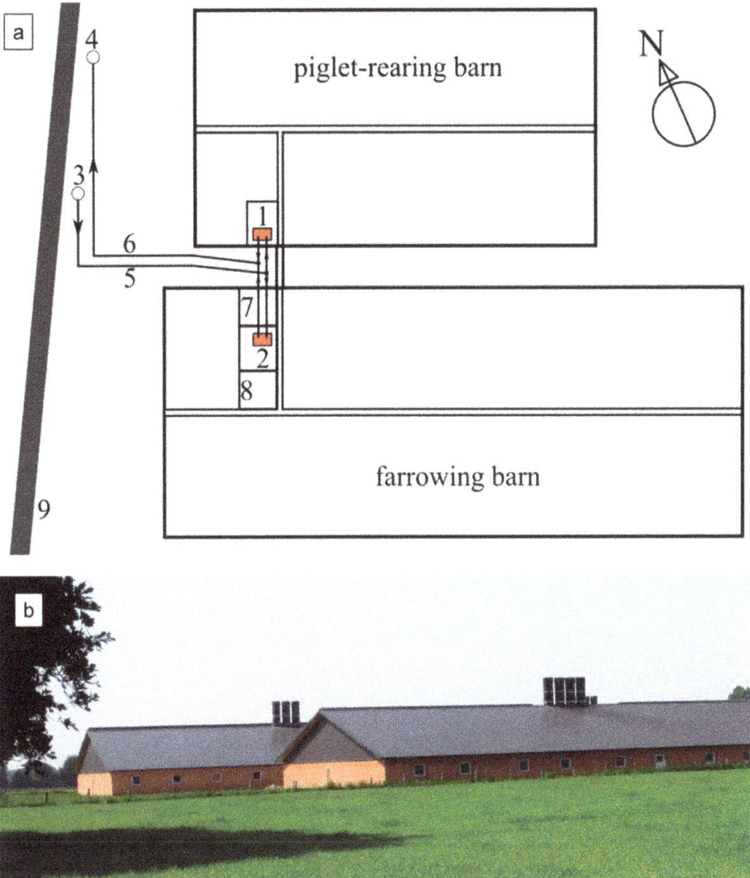

Figure 2. (**a**) Sketch of the floor plan of the barn and (**b**) exterior view of the barn from the north-west side. 1: heat pumps piglet-rearing barn; 2: heat pump farrowing barn; 3: extraction well with groundwater feeding pump; 4: rejection well; 5: groundwater inlet pipe; 6: groundwater outlet pipe; 7: entrance and hygiene lock; 8: barn office; 9: road.

2.1. Description of the Experimental Setup

The farrowing barn consisted of 172 places divided into four compartments. Each compartment had 38 farrowing pens in addition to an extra compartment with 20 farrowing pens. Each pen was equipped with a 0.65 m^2 warm-water piglet heater supplied by the heat pump. For better animal control, no additional thermal insulation covers for the piglet nests were installed. However, additional 100-W infrared heat lamps were used during the new-born piglets' first days. Together, the two modules of the heat pump had a nominal heating capacity of 40 kW. In this barn, modules A and B fed different reservoirs, and module A was responsible for controlling the temperature of the buffer reservoir (500 L) at approximately 45 °C. This serves as a heat source for the warm-water piglet heater. The heat pump of module B was a heat source for the water reservoir (300 L; domestic/service water) and ensured that the water temperature in the reservoir was maintained at approximately 50 °C (Figure 3a).

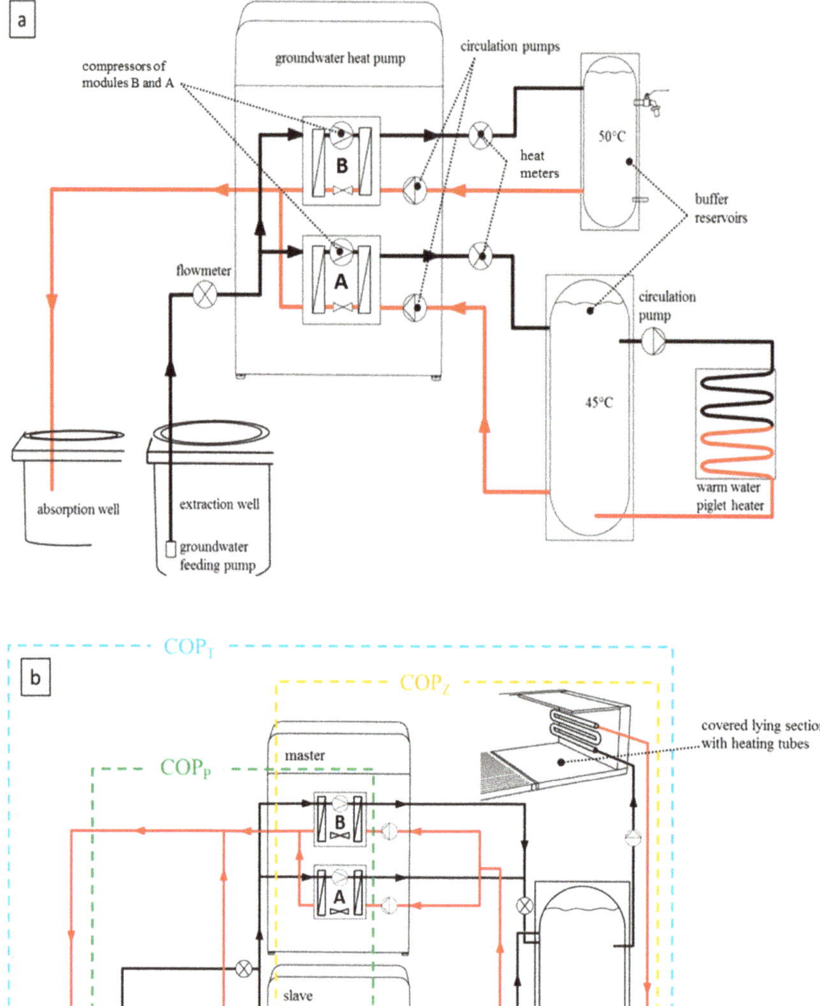

Figure 3. Installation diagram of (**a**) the heat pumps in the farrowing barn with experiment instrumentations and (**b**) piglet-rearing barns with the assessment limits for the determination of the coefficient of performance (COP).

The piglet-rearing barn was divided into eight equal compartments, each with 500 finishing pig places. The compartments were heated by four parallel steel tubes with a diameter of 63 mm. These tubes were mounted on the wall one above the other and the warm water from the heat pump flowed through them. The lying areas of the compartments were provided with heat-protection covers. The compartment temperature depended on the animals' ages and ranged between 32 °C and 24 °C. When the temperature of the compartment decreased by more than 2 °C below the setpoint temperature, an additional 10-kW liquid petroleum gas (LPG) hot air blower turned on automatically. Before restocking the compartments with new animals or to dry the compartments after a cleaning process, 30-kW hot air blowers heated the compartments. Two GWHPs were regularly used to heat the compartments. These pumps were installed in master-slave combination modules; accordingly, the master heat pump operates separately with low heating capacity requirements. The slave heat pump was controlled by the master heat pump and operated only in conditions requiring high heating. All four modules of the two heat pumps were used to supply a 1500-L buffer reservoir with a target temperature of 50 °C (Figure 3b).

A groundwater temperature of approximately 10 °C served as a heat source for both heating circuits of the farrowing and piglet-rearing barns. A frequency-controlled groundwater pump (7.5 kW max 40 m^3 h^{-1}) extracted the groundwater from common extraction well with a depth of 110 m. The groundwater circulated with an operating pressure of 1.3 bar through the heat pump and transferred its heat energy to the working equipment. Once the water passed through the heat pump, it was pumped again into the rejection well, which was 70 m deep and located 20 m away from the extraction well (Figures 2a and 3). The GWHPs of the farrowing and piglet-weaning barn operated independently. Consequently, these two systems can be directly compared.

2.2. Experiment Instrumentations

Four calibrated ALLMESS ultrasonic heat meters (CF Echo II, ALLMESS GmbH, Oldenburg, Germany) were used to measure the heat generated from the groundwater in kilowatt hours (kWh). These sensors had a measuring range for water from 0 to 180 °C with an accuracy of ±5%. To estimate the total amount of water passed through the system, three calibrated flowmeters were used to observe both the groundwater flow for the heat pump systems and the drinking water supply network. Recording both the total amount of water and the water supply of the investigated stables was necessary because both were driven to the same groundwater-feeding pump. In parallel, the energy requirements of the groundwater-feeding pump were recorded using a three-phase meter, to which each groundwater consumer (farrowing or piglet-weaning barn) could be subsequently quantitatively assigned. To achieve energetic comparability of the heat pump systems, an additional electric energy requirement for both pumps was separately detected using three-phase meters. The LPG consumption of the piglet-rearing barn (in m^3) was determined using a calibrated liquid counter (RS/2001AL, Samgas, Vernate Milano, Italy). The outside and inside temperatures of both stables were recorded using temperature data loggers (Testo T175 T1 AG, Lenzkirchen, Germany) with integrated negative temperature coefficient thermistors as sensors. The range of measurement of the sensors was from −35 to +70 °C within ±0.5 °C accuracy.

A special heat pump software program (NIBE, Systemtechnik) was used to record several operating parameters continuously, including the working hours for each heat pump and the input and output groundwater temperatures of the heat pump in degrees Celsius.

2.3. Assessment Limits and COP Calculation

In practice, the efficiency of the heat pump is expressed by the COP (see Equation (1)). Therefore, various possible assessment limits at barn level were used to consider the number of used electrical loads in the system to determine the different efficiency levels of a heating circuit and to carry out weak point analysis of the whole heating circuit. Accordingly, three assessment limits are defined and represented in Figure 3b.

The assessment limits vary depending on the considered number of electrical consumers (Figure 3b). The three assessment limits COP_T (Equation (2)), COP_P (Equation (3)), and COP_Z (Equation (4)) were calculated. They are explained in detail as follows:

$$COP_T = \frac{Q_{HP}}{P_{GFP} + P_{CO,A+B} + P_{CI,A+B} + P_{CI,FB \text{ or } CI,RB}} \quad (2)$$

with

- COP_T: coefficient of performance of the total heating circuit;
- Q_{HP}: thermal heating energy in kWh;
- P_{GFP}: electrical expenditure of the groundwater feeding pump in kWh;
- $P_{CO,A+B}$: electrical expenditure of compressors A + B (in piglet rearing for both heat pumps) in kWh;
- $P_{CI,A+B}$: electrical expenditure of circulation pumps A + B of the heat pump in kWh;
- $P_{CI,FB \text{ or } CI,RB}$: electrical expenditure of the circulation pump for the farrowing or piglet-rearing barns in kWh.

$$COP_P = \frac{Q_{HP}}{P_{GFP} + P_{CO,A+B}} \quad (3)$$

with

- COP_P: coefficient of performance of the groundwater feeding pump and the electrical expenditure of compressors A + B (in piglet rearing for both heat pumps);
- Q_{HP}: thermal heating energy in kWh;
- P_{GFP}: electrical expenditure of the groundwater feeding pump in kWh;
- $P_{CO,A+B}$: electrical expenditure of compressors A + B (in piglet rearing for both heat pumps) in kWh;

$$COP_Z = \frac{Q_{HP}}{P_{CO,A+B} + P_{CI,A+B} + P_{CI,FB \text{ or } CI,RB}} \quad (4)$$

with

- COP_Z: coefficient of performance of the total heating circuit without the electrical expenditure of groundwater feeding pump;
- Q_{HP}: thermal heating energy in kWh;
- $P_{CO,A+B}$: electrical expenditure of compressors A + B (in piglet rearing for both heat pumps) in kWh;
- $P_{CI,A+B}$: electrical expenditure of circulation pumps A + B of the heat pump in kWh;
- $P_{CI,FB \text{ or } CI,RB}$: electrical expenditure of the circulation pump for the farrowing or piglet-rearing barns in kWh.

2.4. Data Analysis

The data collected during the entire investigation period were analyzed using Microsoft Excel 2010. In this descriptive representation the study focused arithmetic mean $\overline{x}$, median $\widetilde{x}$, and the maximum performance potential of the heat pumps under winter conditions.

3. Results

3.1. Climatic Conditions during the Experimental Periods

The average outside air temperature at the experimental period was 2.0 °C ($\widetilde{x}$ = 3.5 °C) (Figure 4). The maximum measured value was 12.8 °C, and the minimum was −15.2 °C. The air temperature in the farrowing barn was nearly constant and averaged 23.0 °C ($\tilde{x}$ = 23.1 °C) (Figure 4a). Whereas the maximum air temperature in the piglet-rearing barn was 34.0 °C and at the end of the rearing period, the compartment air temperature was continuously decreased to 25.0 °C (Figure 4b). The averaged air temperature was 28.4 °C ($\tilde{x}$ = 28.1 °C). The temperature drops in Figure 4 each indicate that the barn is empty.

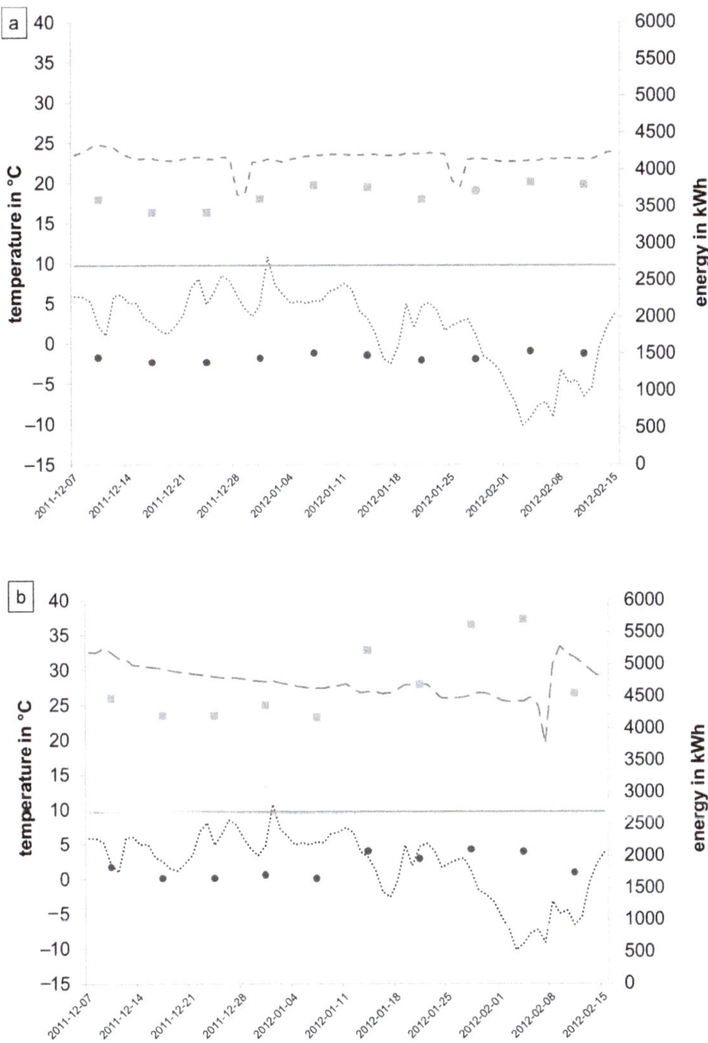

Figure 4. Temperature profiles (hourly mean values) as well as performance of the heat pump(s) (weekly mean values) in (**a**) the farrowing bran and (**b**) the piglet-rearing barn during the trial period. Dotted black line: outside air temperature; dashed grey line: barn inside air temperature; continuous grey line: groundwater temperature; black spots: electrical power consumption of heat pump(s); grey rectangles: heat supplied by the heat pump(s).

3.2. Utilization of Groundwater Heat Pumps

The software of the GWHP read the operating hours of compressors A and B and stored them in the internal data memory. The cumulative data are presented in Figure 5.

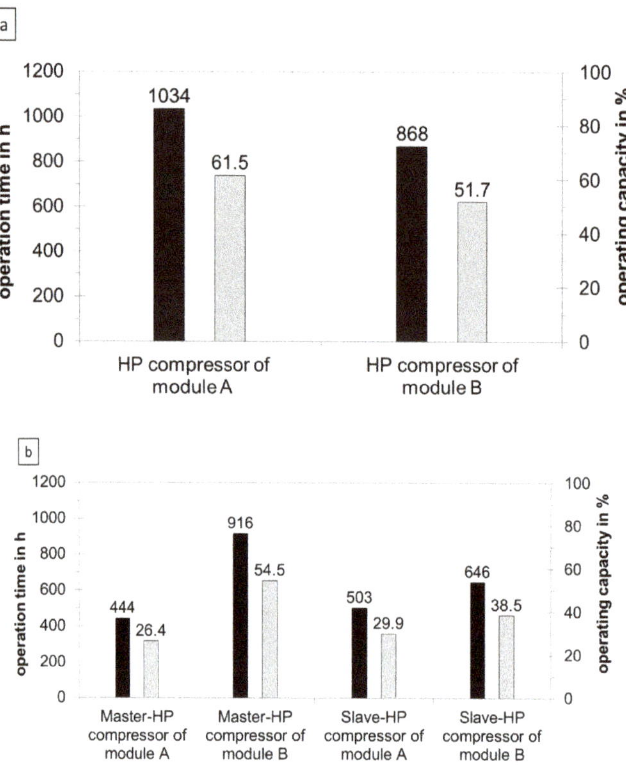

Figure 5. Operating hours (black column) and operating capacity (grey column) of (**a**) heat pump compressors in the farrowing barn and (**b**) piglet-rearing barn during the 70-day trial period, which correspond 1680 h.

In addition to the absolute operating hours of the compressors, the relative operating capacities of the compressors were also specified. During the experimental period of 70 days (cf. Section 2), the operating hours of compressor A in the farrowing barn totaled 1034 h of the 1680 possible hours, which corresponds to an operating capacity of 61.5%. Compressor B reached 868 operating hours with an operating capacity of 51.7% (cf. Figure 5a).

The four heat pump compressors were operated for a total of 2509 h in the piglet-rearing barn during the investigation period, which corresponds to an average operating capacity of 37.3% for the whole system. The compressors of the master heat pump were operated 211 h longer than the cumulative operating times of the two modules (compressors) of the slave heat pump. The highest operating capacity (54.5%) was observed at compressor B of the master module with 916 operating hours (Figure 5b).

3.3. Coefficient of Performance (COP) of the Groundwater Heating Pumps

Figure 4 shows the temperature profiles of groundwater, outside air and inside air as well as the consumption of electric power (input) and the thermal power generated by the heat pumps (output) during the trial period for the farrowing barn (Figure 4a) and the piglet-rearing barn (Figure 4b). Both figures show in different ways that when outside air temperatures decrease, the power consumption of the heat pumps increases as well as the thermal output of the heat pumps, and vice versa.

Figure 6 shows the electric energy consumption considering the three assessment limits (see Equations (2)–(4) and Figure 3b) and the generated usable heat energy for both the farrowing barn (Figure 6a) and the piglet-rearing barn (Figure 6b).

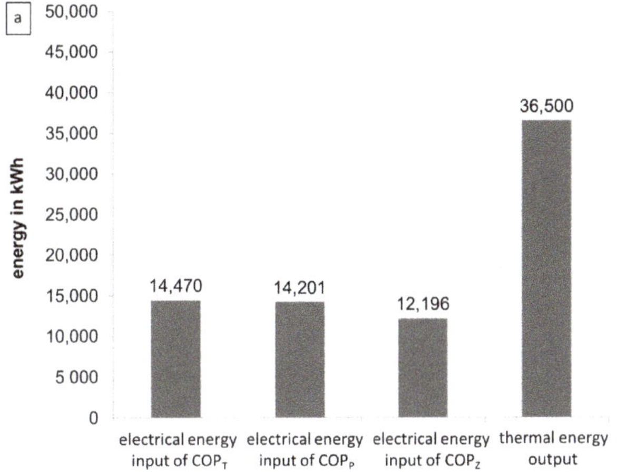

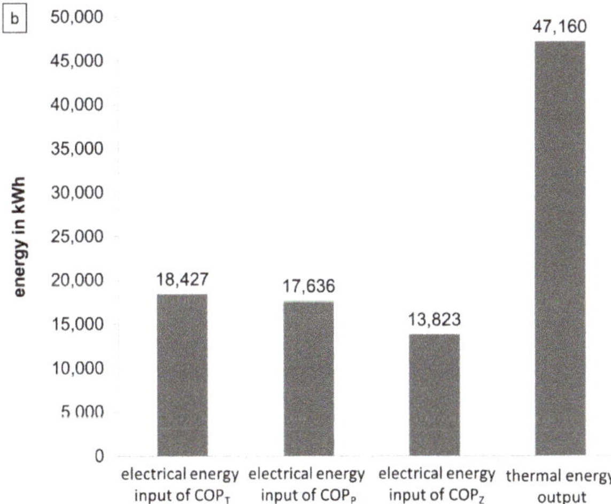

Figure 6. Electrical energy input depended on the different assessment limits (see Section 2.3) and thermal energy output during the experimental period in (**a**) the farrowing barn and in (**b**) the piglet-rearing barn.

The total consumed energy during the whole measuring period of the heat pump and the groundwater feeding pump (GFP) in the farrowing barn was 14,201 kWh (Figure 6a). During the same period, the heat pump provided 36,500 kWh of total utilized thermal energy. Accordingly, the COP_P value reached 2.6. After adding the electricity consumption of 269 kWh_{el} by the three circulation pumps for the heating circuits the COP_T decreased to 2.5. Considering the overall system without the electricity consumption of the GFP resulted in a COP_Z of 3.0 (cf. Figure 6a and Table 1).

Table 1. Comparison of two heat pump systems of two studies.

Heat Pump System	GWHP [cf. 36]	GWHP [cf. 36]	Brine heat Pump with Area Collectors [29]	Brine Heat Pump with Area Collectors [29]
Source	Groundwater	Groundwater	Ground	Ground
Types of livestock housing	Piglet-rearing barn	Farrowing barn	Piglet-rearing barn	Farrowing barn
Experimental period in d	70	70	146	364
Electricity consumption in kWh_{el}				
- HP with circulation pumps	13,823	12,196	6725	8000
- HP with GFP or brine pump	17,636	14,201	8732	-
- HP with circulation pumps and GFP or brine pump	18,427	14,470	9961	-
Σ Heat output in kWh_{th}	47,160	36,500	25,467	32,000
COP_Z	3.4	3	3.8	4
COP_P	2.7	2.6	2.6	-
COP_T	2.6	2.5	2.6	-

For the master-slave heat pump in the piglet-rearing barn, 17,636 kWh_{el} was consumed by the heat pump and the GFP (Figure 6b). For heating over the same period, the heat pump provided 47,160 kWh_{th} of utilizable heat. The COP of the heat pump was calculated similarly to the COP of the heat pump in the farrowing barn using Equations (2)–(4). The calculated values for COP_P and COP_T were 2.7 and 2.6, respectively. However, the COP_Z value was 3.4 for the heat pump system without considering the GFP (cf. Figure 6b and Table 1).

3.4. Heat Quantity and Produced Heating Capacity

A complement of the required heating capacity of the heat pump was represented to provide the necessary heat quantity. A total of 36,160 kWh of thermal energy provided by the heat pump was used to heat the warm water flowing through the piglet nests (farrowing barn). Therefore, each piglet nest used 3.0 kWh daily in winter. During the experimental period, an average of 23 kW of power was retrieved, representing approximately 57.5% of the maximum heat pump manufacturer's specifications.

The piglet-rearing barn was supplied with a total of 55,216 kWh_{th} using a hot air blower and GWHP with a share of 47,160 kWh_{th} during the experimental period. The heat pump provided 85.4% of thermal energy, whereas the hot air blower produced 14.6%. The average equivalent value was approximately 790 kWh per day or 0.2 kWh per fattening place and day. The required heating capacity for the piglet-rearing barn was calculated by considering the heat pump and hot air blower as an average of 37 kW (σ = 7.9 kW; $\widetilde{x}$ = 36.1 kW). This value corresponds to an average of 46.3% of the maximum system performance for the master-slave configuration.

3.5. Groundwater Flow Rates

For the heat pump applications in the farrowing and piglet-rearing barns, approximately 18,049 m^3 of groundwater was obtained from the extraction well during the experimental period. This groundwater passed through the heat pump and then fed into the rejection well. This equivalent daily groundwater requirement is specified under the combined operating loads in Figure 5a,b as 258 m^3. The groundwater enters at a constant supply temperature of $\bar{x}$ = 9.7 °C (σ = 0.1 °C; $\tilde{x}$ = 9.7 °C) in the heat pump (Figure 4) and leaves it with an average temperature of 6.7 °C (σ = 1.4 °C; $\tilde{x}$ = 6.7 °C). The groundwater was consequently cooled in the heat pump by an average of 3.0 °C. Approximately 3.5 kWh_{th} m^{-3} of thermal energy is used to process the groundwater through the heat pump on average.

The water quantity is specified as shown in Figure 7 and was promoted daily by the heat pumps and the ratio between the daily amount of produced heat and the daily groundwater flow through the systems.

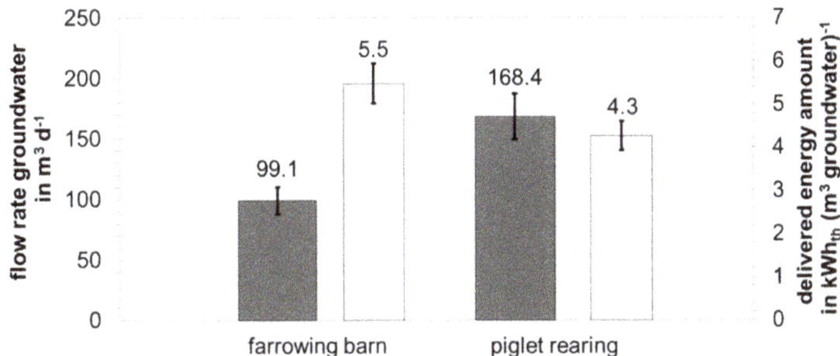

Figure 7. Daily groundwater flow rate (dark column) through the heat pumps and the achieved thermal energy per m^3 groundwater (light column). The equivalent standard deviations are also presented.

The farrowing barn was provided with purposed heat using the heat pump with an average of 5.5 kWh_{th} for each cubic meter of groundwater (σ = 0.5 kWh_{th} m^{-3}; $\tilde{x}$ = 5.5 kWh_{th} m^{-3}). The piglet-rearing barn was provided with average purposed heat of 4.3 kWh_{th} m^{-3} (σ = 0.3 kWh_{th} m^{-3}; $\tilde{x}$ = 4.3 kWh_{th} m^{-3}) via 168.4 m^3 of daily groundwater flow.

4. Discussion

4.1. Climatic Conditions during the Experimental Periods

The climatic weather conditions at the experimental site represent a classic winter in this region. The inside temperatures of the farrowing barn and piglet-rearing barn (cf. Figure 4) shows a constant and good climate during the test period. The barn inside temperatures correspond to the requirements described in the introduction.

4.2. Groundwater Heat Pumps

The degree of utilization of the heat pumps was relatively low because the dimensions of the investigated heat pumps were subsequently doubled for both stables. By doubling the size of the farrowing barn, a power deficit in the heat pump occurred with a nominal power of 40 kW, which could be balanced using hot air blowers. The produced heat output of the heat pump in the piglet-rearing barn was as expected from the sizing of the system. No additional costs will be incurred by the future extension of the same piglet-rearing barn construction after the heat transfer system establishing in the compartments. In this case, the nominal power will be 80 kW.

As described in the introduction, German authorities must approve the extraction and recirculation of the groundwater when a GWHP is used. On this experimental site, the approval by the German Water Act went without a problem. The groundwater quality fully met the legal requirements. Manganese or iron deposits on the exchanger surfaces were not found by inspecting the barn facilities.

During the experimental period, the extraction well had a daily average groundwater quantity of 99.1 m^3 for the heat pump circuit in the farrowing barn and 168.4 m^3 d^{-1} in the piglet-rearing barns (Section 3.5.; Figure 7). An average of 69.3 m^3 d^{-1} more groundwater was pumped through the system in the piglet-rearing barn than in the farrowing barn, corresponding to an additional groundwater consumption of 69.9%. It is mentioned by Matthias [17] that the groundwater flow capacity required for the operation of a GWHP is 48 m^3 d^{-1}. The measured values exceeded the specified value in the literature by a factor of 2.1 to 3.5. However, the investigated GWHPs achieved only approximately 50% of their theoretical capacities. Accordingly, the extracted groundwater requirements should increase at full capacity.

Regarding the required groundwater circuit capacity for the extraction well, it should be noted that this capacity depends on the system performance, the type of application, and the choice of groundwater flow rate. Therefore, fixed guide values cannot be provided at this time.

Cremer [29] estimates the possibility of using heat pumps in livestock buildings because of the relatively low temperature supply (approx. 45 °C) when low-temperature warm-water heaters are used efficiently and, thus, economically. The piglet nests in the farrowing barn were operated with a warm-water supply temperature of 45 °C. Only a very small portion of the produced heat (0.9%) was required to heat the 50 °C domestic water. The supply temperature of the heating circuit mentioned by Cremer [29] was exceeded by 5 °C in the investigated piglet-rearing barn. The observed temperature accounted for 50 °C. Due to the increased supply temperature, negative effects are already expected in the development of the COP for the heat pump system.

4.3. Comparison with Other Heat Pumps Systems

The COP values achieved were between 2.5 and 3.4. The price development for primary energy must be considered when comprehensively evaluating this system. By increasing fossil energy prices, the heat pump can be economically operated, even at COP values less than 4.0, compared to conventional fossil heating. However, the value 4.0 fluctuates depending on price of fossil heating energy costs.

To improve the COP of the heat pump, the groundwater flow was increased and the supply temperature in the farrowing barn decreased. Accordingly, the temperature difference between the heat source and heat sink was reduced, thereby improving the COP. This modification was halted after a short time because the desired lying area temperature of approximately 38–40 °C in the farrowing barn was no longer reached in the farrowing barn due to the lower supply temperature.

The COP serves as the base value for comparing heat pumps (see Section 1). Due to a lack of literature references relating to the use of heat pumps in livestock stables, a direct comparison can only be made to the research of Cremer [29], considering the assessment limits (Figure 3b). Table 1 shows that COP_P and COP_T in the present study are largely identical to those of the study of Cremer [29] (investigation of a piglet-rearing barn and farrowing barn with brine heat pump with area collectors). Furthermore, it becomes clear that the COP_Z in both studies are significantly higher than COP_T and COP_P. This effect is due to the electricity consumption by the GFP (in this study) and the brine pump for the area collectors (at Cremer [29]). Furthermore, the heat transport from the heat pump into the barn itself is decisive for the COP value. Cremer investigated an outdoor climate piglet rearing house (1100 animal places) in which the heat pump was used to heat cubicles with underfloor heating [29]. In the piglet-rearing barn of this study, the heat in the compartments is distributed via four parallel steel tubes as space heating (see Section 2.1). This requires a higher water circulation; the flow temperature is negatively influenced and the temperature difference between sink and source is increased. These factors also influence the efficiency of the entire heating system, but to a lesser extent than the feed pump for the groundwater.

The shares of electrical consumers in total electricity consumption in both studies are comparable (Table 1): the highest share is due to the operation of the heat pump itself, followed by the pumping technology for the medium (from the sink to the heat pump). The smallest share of electrical energy is used for the distribution of heat in the stable system.

The evaluation of the COP varies. While Lucia et al. [19] report that the COP of ground-source heat pumps ranges from 3.0 to 3.8, Matthias [17] points out that a COP greater than 4.0 should be targeted for heat pumps in order to ensure economical operation. It should be noted that the COP_T number is largely dependent on the dimensioning of the overall system (heat pump, circulating or feed pump, circulating pumps for heat transfer into the building, and the technique for heat dissipation) and requires careful dimensioning.

4.4. Substitution Potential Power of Fossil Energy and a Contribution to CO_2 Savings

A reduction in CO_2 emissions is also important for achieving the defined climate protection targets (cf. Section 1). Agricultural livestock farming can contribute to this by using renewable energies. By using heat pumps in this study, 47,160 kWh (piglet-rearing barn) and 36,500 kWh (farrowing barn) of thermal energy could be generated (Section 3.3). If this amount of heat had to be generated with fossil fuels, 4716 L of fuel oil or 7178 L of liquid gas would have been required for the piglet house, and 3650 L of fuel oil or 5556 L of liquid gas for the farrowing house. Consequently, the use of heat pumps in this experiment avoided 8.5–12.5 tons of CO_2 emissions. From an environmental point of view, this is assessed positively.

The avoidance of CO_2 emissions by using heat pumps is confirmed by Islam et al. [33]. These authors also report that in their experiment other emissions (ammonia, hydrogen sulfide, sulfur dioxide) besides CO_2 were also significantly reduced.

4.5. Groundwater Heat Pump Compared to Oil or Gas Heating

It is not easy to make a direct comparison between the groundwater heat pump and an oil or gas heating system, since any comparison is very much dependent on the respective calculation bases and the energy prices, which, as is well known, can fluctuate very strongly. Nevertheless, a rough calculation was made. The following systems were placed side by side: a groundwater heat pump with a COP_T of 2.7, a groundwater heat pump with a COP_T of 3.5, a natural gas heating system, and an oil heating system. The nominal heating capacity of all heating systems is 120 kW each. The annual heating requirement is estimated at 400,000 kWh. The electric power is always taken into account with a net electricity price of €0.18 kWh^{-1}. The acquisition costs for the heat pump are estimated at €44,300 and for the oil and natural gas heating each at €17,000. Interest rate (3.5%) and depreciation over 10 years (incl. dept service) as well as maintenance cost are considered. The only variable factor in the calculation is the energy prices for oil and natural gas over the last 10 years (annual average values). This result of this calculation is shown graphically in Figure 8. For better understanding, trend lines were inserted.

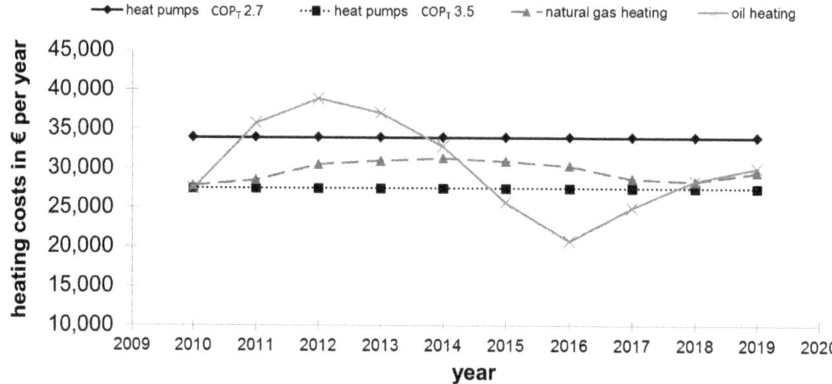

Figure 8. A comparison of the annual heating costs for four different heating systems (a groundwater heat pump with a COP_T of 2.7, a groundwater heat pump with a COP_T of 3.5, a natural gas heating system and an oil heating system) over the last 10 years. The only variable factor is the energy price (annual average) for oil and natural gas.

The graphs of the heat pumps in Figure 8 run parallel to the abscissa, since the heat pumps are operated independently of oil or natural gas. However, the influence of the COP is very clear. With a COP of 3.5, the performance ratio is significantly better, which is also associated with significantly lower annual costs. There is a clear link between the annual costs of natural gas and oil heating and energy prices, although the oil price usually fluctuates more strongly than the natural gas price. In the years 2011 to 2013, heat pump operation is even more economical than oil heating, even with a poor COP of 2.7, so the farm owner's decision to use heat pump technology in 2009 is understandable in view of rising energy prices (at that time). In the following years, on the other hand, operation with an oil heating system would have been cheaper than with a heat pump, because the oil price has fallen relatively sharply again.

The operation of a natural gas heating system is a sensible alternative if a sufficiently high COP number for heat pump operation cannot be guaranteed. However, it must be considered that natural gas is not available in all areas.

If the price of electricity continues to rise and natural gas and oil prices should stagnate, a heat pump is not recommended under current conditions.

5. Conclusions

The present study of two heat pump systems contributes to the practical analysis of the performance of regenerative heating technologies. Heat pumps can be characterized realistically in terms of the relationship between electricity consumption and both energy production and energy efficiency. The economic and ecological operation of a groundwater heat pump depends on achieving a high coefficient of performance. For this purpose, the lowest possible flow temperatures of the compartment heating circuit should be selected, as this is the only way to ensure a small temperature difference between the heat source (ground water) and the heat sink (heating circuit flow) at a relatively constant ground water temperature. For this reason, the installation of a groundwater heat pump, it is essential to adjust the heating regime in the barn compartments to low heating circuit flow temperatures.

The following aspects should be considered when installing a GWHP in a farrowing barn or a piglet rearing barn:

- The warm-water piglet heater is preferable to the space heating variants due to the lower flow temperature.

- Hot water pipes of the heating circuit should be thermally insulated. As a result, the transport heat losses remain low and the flow temperatures can be selected lower, correspondingly.
- In addition, the warm-water piglet heater should be routed in a parallel proceeding to ensure a homogeneous heat distribution in the common area of the animals.
- The radiating surface of the heating elements in the animal house must be sufficiently large.
- The lying area should be provided with a heat protection cover. As a result, the animals can be provided with an optimal microclimate even at low heating circuit flow temperatures.

The above-mentioned installation recommendations represent an opportunity to minimize flow temperature. This describes the most important and influenceable regulating variable for an economic and ecological operation of a GWHP.

In addition, the following points can be noted:

- The permanently constant ground water temperature has a positive effect on the performance of a heat pump.
- However, pumping groundwater requires a considerable amount of electrical power by the groundwater feeding pump.
- The heat pump was oversized for the stable complex under investigation. (Background: the barn should be doubled at a later date. However, the heat pumps were dimensioned directly for the entire building.)
- The application of the three different assessment limits could help to identify factors influencing the coefficient of performance.
- The high supply temperature of 50 °C in the piglet rearing barn has a negative effect on the COP. It is recommended that the radiating area in the house is increased (up to now only steel tubes have been installed here), so that the required heat output in the house can still be ensured at low supply temperature.
- The data collected refer to the winter.
- Against the background of the sharp drop in feed-in tariffs for electricity from photovoltaic systems in Germany, it should be examined whether it makes sense to install a photovoltaic system on the roof surfaces to generate electricity. On the one hand, the use of self-generated electricity could lead to becoming somewhat more independent of the energy market. On the other hand, the operation of the heat pump could become more economical if the kilowatt hour of electrical power generated on the own farm is cheaper than the kilowatt hour purchased from the network provider.

It would be desirable if such a heat pump system were examined over a whole year in order to obtain a holistic picture of the groundwater heat pump and its performance over all seasons. It would also be interesting to check whether a combination of heat pumps with other house systems makes sense. In this connection, a modular housing system with geothermal heat exchanger, for example, could be combined with a brine heat pump. The brine circuit of the heat pump could be integrated into the central exhaust air collecting duct of the modular house. The exhaust air could thus be extracted at a high temperature. This installation would have the advantage of a lower temperature difference between the heat source (warm exhaust air) and heat sink (heating circuit), which should result in a higher COP. However, a more secure and continuous outflow in the exhaust air-collecting duct of the condensate should be ensured to prevent building damage. However, other solutions are also conceivable. A very important future research question in times of global warming is to what extent heat pump systems can be used not only for heating in winter but also for cooling in summer (cf. [32,34,35]) and which modifications are associated with this.

Author Contributions: Conceptualization, H.L., P.R., M.S.K. and W.B.; Methodology, H.L., P.R. and W.B.; Software, H.L. and P.R.; Validation, H.L., P.R. and M.S.K.; Formal Analysis, H.L. and P.R.; Investigation, H.L. and P.R.; Resources, W.B.; Data Curation, H.L., P.R. and M.S.K.; Writing—Original Draft Preparation, H.L. and M.S.K.;

Writing—Review and Editing, H.L., M.S.K., E.M. and W.B.; Visualization, H.L. and P.R.; Supervision, M.S.K. and W.B.; Project Administration, W.B.; Funding Acquisition, P.R. and W.B. All authors have read and agreed to the published version of the manuscript.

Funding: This research was founded by the Lehr- und Forschungsschwerpunkt "Umweltverträgliche und Standortgerechte Landwirtschaft" (USL).

Acknowledgments: The authors thank the farm owner for the permission to carry out the experiments on his farm. They would also like to thank Gerd-Christian Maack (Institute of Agricultural Engineering, University of Bonn) for his assistance and discussion of the results and for preparing Figure 8, and Hauke F. Deeken (Institute of Agricultural Engineering, University of Bonn) for preparing Figure 2a.

Conflicts of Interest: The authors declare no conflict of interest.

Abbreviations and Indices

COP	coefficient of performance
GFP	groundwater feeding pump
GWHP	groundwater heat pump
P	electric power consumption
Q	thermal energy in kWh
$\dot{Q}$	heating performance in kW
CI,A+B	circulation pump A and B
CO,A+B	compressor A and B
el	electrical
FB	farrowing barn
HP	heat pump
RB	piglet-rearing barn
T	total electric power consumption of all compressors and circulation pumps of the heating system
th	thermal
Z	the electric power consumption of the circulation pumps A + B of the heat pump (in piglet rearing with both heat pumps) and the electric power consumption of the circulation pumps for the farrowing and piglet-rearing barns

References

1. German Environment Agency. Climate Footprint 2018: 4.5 Percent Decrease in Greenhouse Gas Emissions. German Environment Agency Produces Initial Detailed Estimate. 2019. Available online: https://www.umweltbundesamt.de/en/press/pressinformation/climate-footprint-2018-45-percent-decrease-in (accessed on 15 January 2020).
2. Corré, W.; Schröder, J.; Verhagen, J. Energy use in conventional and organic farming systems. In Proceedings of the Open Meeting of the International Fertiliser Society, London, UK, 3 April 2003; London International Fertiliser Society: York, UK, 2003. ISBN 0-85310-147-7. ISSN 1466-1314.
3. Krommweh, M.S.; Rösmann, P.; Büscher, W. Investigation of heating and cooling potential of a modular housing system for fattening pigs with integrated geothermal heat exchanger. *Biosyst. Eng.* **2014**, *121*, 118–129. [CrossRef]
4. Büscher, W. Energiebedarf und Leistungsbereitstellung in der Schweinehaltung. In *Baubriefe Landwirtschaft e. V. 47*; Landwirtschaftsverlag GmbH: Münster-Hiltrup, Germany, 2009; pp. 10–15.
5. Heinritzi, K.; Gindele, H.R.; Reiner, G.; Schnurrbusch, U. Allgemeiner Untersuchungsgang. In *Schweinekrankheiten, UTB 8325*; Verlag Eugen Ulmer KG: Stuttgart, Germany, 2006; pp. 18–22.
6. European Commission. *Best Available Techniques (BAT) Reference Document for the Intensive Rearing of Poultry or Pigs, Industrial Emissions Directive 2010/75/EU Integrated Pollution Prevention and Control*; EU 28674 EN; Publication Office of the European Union: Luxemburg, 2017. [CrossRef]
7. Meyer, E.; Vogel, M.; Wähner, M. Investigations on acceptance and size of piglet nests. *Landtechnik* **2012**, *67*, 362–365.

8. DIN 18910:2017-08. *Thermal Insulation for Closed Livestock Buildings—Thermal Insulation and Ventilation—Principles for Planning and Design for Closed Ventilated Livestock Buildings*; DIN Deutsches Institut für Normung e. V., Ed.; Beuth Verlag GmbH: Berlin, Germany, 2017.
9. Büscher, W.; Cremer, P.; Feller, B.; Fritzsche, S. *Lüftung und Wärmedämmung geschlossener Ställe—Bemessung nach DIN 18910:2017-08*; Sonderveröffentlichung, Ed.; Kuratorium für Technik und Bauwesen in der Landwirtschaft e. V. (KTBL): Darmstadt, Germany, 2018; ISBN 978-3-945088-61-6.
10. Lindley, J.A.; Whitaker, J.H. *Agricultural Buildings and Structures*; ASAE (The Society for Engineering in Agricultural, Food, and Biological Systems): St. Joseph, MI, USA, 1996; pp. 249–270, 347–356. ISBN 978-0-929355-73-3.
11. Spengler, R.W.; Stombaugh, D.P. Optimization of earth-tube heat exchangers for winter ventilation of swine housing. *Trans. ASAE* **1983**, *26*, 1186–1193. [CrossRef]
12. Van Caenegem, L. *Energieeffizienz in Abferkelställen Durch Erdwärmenutzung, Energieeffiziente Landwirtschaft*; KTBL-Schrift 463; Kuratorium für Technik und Bauwesen in der Landwirtschaft e. V. (KTBL): Darmstadt, Germany, 2008; pp. 162–172. ISBN 978-3-939371-59-5.
13. Rösmann, P.; Büscher, W. Rating of an air-to-air heat exchanger in practice. *Landtechnik* **2010**, *65*, 418–420. [CrossRef]
14. Allen, W.H.; Payne, F.A. Designing animal ventilation schedules with counterflow heat exchangers. *Trans. ASAE* **1987**, *30*, 782–788. [CrossRef]
15. VDI 3803-5. *VDI-Richtlinie 3803-5: Air-Conditioning, System Requirements. Part 5: Heat Recovery Systems*; Verein Deutscher Ingenieure e. V. (VDI), Ed.; Beuth Verlag GmbH: Berlin, Germany, 2013.
16. Recknagel, H.; Sprenger, E.; Albers, K.J. *Taschenbuch für Heizung und Klimatechnik, Band 1*; Deutscher Industrieverlag GmbH (DIV): München, Germany, 2015; p. 1035. ISBN 978-38356-7136-2.
17. Matthias, J. Wärmepumpen zur Stallbeheizung. In *Baubriefe Landwirtschaft e. V. 47*; Landwirtschaftsverlag GmbH: Münster-Hiltrup, Germany, 2009; pp. 70–72. ISBN 978-3-7843-3410-3.
18. Baumann, M.; Laue, H.-J.; Müller, P. *Wärmepumpen—Heizen mit Umweltenergie, 4. Erweiterte und Vollständig Überarbeitete Auflage, BINE Informationsdienst*; Verlag Solarpraxis AG: Berlin, Germany, 2007; ISBN 978-3-934595-60-6.
19. Lucia, U.; Simonetti, M.; Chiesa, G.; Grisolia, G. Ground-source pump system for heating and cooling: Review and thermodynamic approach. *Renew. Sustain. Energy Rev.* **2017**, *70*, 867–874. [CrossRef]
20. VDI 4645:2018-03. *VDI-Guideline 4645: Heating Plants with Heat Pumps In Single-Family and Multi-Family Houses, Planning, Construction, Operation*; Verein Deutscher Ingenieure e. V. (VDI), Ed.; Beuth Verlag GmbH: Berlin, Germany, 2018.
21. Omer, A.M. Ground-source heat pumps systems and applications. *Renew. Sustain. Energy Rev.* **2008**, *12*, 344–371. [CrossRef]
22. Ahmadi, M.H.; Ahmadi, M.A.; Sadaghiani, M.S.; Ghazvini, M.; Shahriar, S.; Nazari, M.A. Ground source heat pump carbon emissions and ground-source heat pump systems for heating and cooling of buildings: A review. *Environ. Progress Sustain. Energy* **2017**, *37*, 1241–1265. [CrossRef]
23. Tissen, C.; Menberg, K.; Bayer, P.; Blum, P. Meeting the demand: Geothermal heat supply rates for an urban quarter in Germany. *Geother. Energy* **2019**, *7*, 9. [CrossRef]
24. Zurmühlen, S.; Wolisz, H.; Angenendt, G.; Magnor, D.; Streblow, R.; Müller, D.; Sauer, D.U. Potential and Optimal Sizing of Combined Heat and Electrical Storage in Private Households. *Energy Procedia* **2016**, *99*, 174–181. [CrossRef]
25. Meng, B.; Vienken, T.; Kolditz, O.; Shao, H. Modeling the groundwater temperature response to extensive operation of ground source heat pump systems: A case study in Germany. *Energy Procedia* **2018**, *152*, 971–977. [CrossRef]
26. Ziegler, T.; Teodorov, T.; Mellmann, J. Fixed bed drying of medicinal plants using dehumidification of air. *Landtechnik* **2011**, *66*, 167–169. [CrossRef]
27. Benli, H.; Durmus, A. Evaluation of ground-source heat pump combined latent heat storage system performance in greenhouse heating. *Energy Build.* **2009**, *41*, 220–228. [CrossRef]
28. Tong, Y.; Kozai, T.; Nishioka, N.; Ohyama, K. Greenhouse heating using heat pumps with a high coefficient of performance (COP). *Biosyst. Eng.* **2010**, *106*, 405–411. [CrossRef]
29. Cremer, P. *Wärmepumpeneinsatz in Landwirtschaftlichen Betrieben—Hinweise und Beispiele*; Förderkreis Stallklima: Dummerstorf, Germany, 2011.

30. Lücke, W.; Hörsten, D. Utilising Heat in Roofs for Energy Generation in Agricultural Buildings. *Landtechnik* **2006**, *61*, 208–209. [CrossRef]
31. Riva, G.; Pedretti, E.F.; Fabbri, C. Utilization of a Heat Pump in Pig Breeding for Energy Saving and Climate and Ammonia Control. *J. Eng. Res.* **2000**, *77*, 449–455. [CrossRef]
32. Manolakos, D.; Panagakis, P.; Bartzanas, T.; Bouzianas, K. Use of heat pumps in HVAC systems for precise environment control in broiler houses: System´s modeling and calculation of the basic design. *Comput. Electron. Agric.* **2019**, *163*, 104876. [CrossRef]
33. Islam, M.M.; Mun, H.-S.; Rubayet Bostami, A.B.M.; Ahmed, S.T.; Park, K.-J.; Yang, C.-J. Evaluation of a ground source geothermal heat pump to save energy and reduce CO_2 and noxious gas emissions in a pig house. *Energy Build.* **2016**, *111*, 446–454. [CrossRef]
34. Alberti, L.; Antelmi, M.; Angelotti, A.; Formentin, G. Geothermal heat pumps for sustainable farm climatization and field irrigation. *Agric. Water Manag.* **2018**, *195*, 187–200. [CrossRef]
35. Wang, M.Z.; Wu, Z.H.; Chen, Z.H.; Tian, J.H.; Liu, J.J. Economic Performance Study on the Application of Ground Source Pump System in Swine Farms in Beijing China. *AASRI Procedia* **2012**, *2*, 8–13. [CrossRef]
36. Rösmann, P. Einsatz von Regenerativen Energiequellen zum Heizen und Kühlen von Zwangsbelüfteten Tierställen. Ph.D. Thesis, Rheinische Friedrich-Wilhelms-Universität Bonn, Bonn, Germany, 2012.
37. Costantino, A.; Fabrizio, E.; Ghiggini, A.; Bariani, M. Climate control in broiler houses: A thermal model for the calculation of the energy use and indoor environmental conditions. *Energy Build.* **2018**, *169*, 110–126. [CrossRef]

Article

Temperature Measurements on a Solar and Low Enthalpy Geothermal Open-Air Asphalt Surface Platform in a Cold Climate Region

Caner Çuhac, Anne Mäkiranta *, Petri Välisuo, Erkki Hiltunen and Mohammed Elmusrati

School of Technology and Innovations, University of Vaasa, P.O. Box 700, FI-65101 Vaasa, Finland; canercuhac@gmail.com (C.Ç.); petri.valisuo@uva.fi (P.V.); erkki.hiltunen@uva.fi (E.H.); mohammed.elmusrati@uva.fi (M.E.)

* Correspondence: anne.makiranta@univaasa.fi; Tel.:+358-29-449-8302

Received: 10 December 2019; Accepted: 18 February 2020; Published: 21 February 2020

Abstract: Solar heat, already captured by vast asphalt fields in urban areas, is potentially a huge energy resource. The vertical soil temperature profile, i.e., low enthalpy geothermal energy, reveals how efficiently the irradiation is absorbed or radiated back to the atmosphere. Measured solar irradiation, heat flux on the asphalt surface and temperature distribution over a range of depths describe the thermal energy from an asphalt surface down to 10 m depth. In this study, those variables were studied by long-term measurements in an open-air platform in Finland. To compensate the nighttime heat loss, the accumulated heat on the surface should be harvested during the sunny daytime periods. A cumulative heat flux over one year from asphalt to the ground was 70% of the cumulative solar irradiance measured during the same period. However, due to the nighttime heat losses, the net heat flux during 5 day period was only 18% of the irradiance in spring, and was negative during autumn, when the soil was cooling. These preliminary results indicate that certain adaptive heat transfer and storage mechanisms are needed to minimize the loss and turn the asphalt layer into an efficient solar heat collector connected with a seasonal storage system.

Keywords: asphalt solar collector; heat flux; distributed temperature sensing; low enthalpy geothermal energy; renewable energy; soil temperature profile

---

1. Introduction

Today's society is obliged to search for new low carbon energy resources to fight against global climate change. In cold-climate regions, fossil fuels are often used for heating. Solar energy, which is one of the most important inexhaustible sustainable energy resources, can be harvested using heat collectors or solar panels. Solar heat is renewable, and therefore, can potentially reduce dependency on fossil fuels. On the other hand, collecting of solar energy from the asphalt layer during the hottest season also saves the asphalt from large temperature changes, which can cause structural damage like rutting or hardening [1,2].

Surface temperature changes caused by solar irradiation degrade exponentially with a time constant of a few hours due to the thermal energy flowing back into the atmosphere. The accumulation of the daily gains or losses in terms of solar energy leads to small temperature increases in the deeper layers of the soil, creating long-term warming or cooling of the soil [3,4]. A diurnal and/or seasonal storage is necessary for a solar heating system in a cold climate region. Majorowicz et al. [5] and Loomans et al. [6] have expressed an application of an asphalt collector for summer and winter conditions. They have also studied the thermal energy potential of an asphalt collector and what the critical parameters are.

Some simple storage systems operate without heat pumps; only solar collectors and panels are connected with seasonal thermal energy storage [7]. Solar heat collectors, together with heat pumps and seasonal thermal energy storage systems, are frequently used in heating real estate, industrial buildings, and greenhouses [8,9]. Drake Landing Solar Community in Canada has experience in the harvesting and storing of solar energy [10,11]. In Drake Landing, 52 detached houses are heated using 2293 m^2 of flat plate solar collectors. The storage volume used for seasonal storage consisted of 144 boreholes, each being the depth of 35 m. The temperature of the borehole thermal energy storage system (BTES) reached above 65 °C in summer after three years of operation, and the temperature dropped nearly at 40 °C during the winter [10,12]. Experimental measurements and a Comsol simulation model of thermal energy from solar collectors in the ground were studied also by Haq and Hiltunen [13]. The theories and models used by Haq and Hiltunen can later be used to optimize the parameters of the asphalt heat collection and seasonal storage system.

The focus and novelty of this research was to study the thermal energy absorption of asphalt surface and soil layers beneath an asphalt layer in a cold climate region. The goal was reached by benchmarking and analyzing the solar irradiance and the absorption rates of the asphalt surface around the clock in different seasons. The vertical temperature distribution in the soil was studied for possible solar and low enthalpy geothermal energy harvesting applications in future.

The laboratory experiments, carried out before this study, indicated that dark asphalt is efficient in absorbing solar irradiance and conducting thermal energy to the soil [14]. Ho et al. [15] has modeled an asphalt paved area with most important parameters and variables by Comsol finite element simulation. In Finland, logistic centers are interested in snow melting systems. In this research, the thermal behavior of commonly used asphalt paved soil structures, e.g., parking spaces, is studied in high latitudes above 63° N at University of Vaasa campus site (see Figure 1), which is farther north than the previous research projects cited above. The monthly mean air temperatures and heating degree days in Vaasa during the study period are declared in Table 1.

Figure 1. Location of measurement site marked by red circle, Vaasa, Finland [16].

Table 1. The monthly mean air temperatures [°C] measured during April 2014–December 2015 at the weather station of the Finnish Meteorological Institute [17] and the heating degree days (HDD) in Vaasa [18].

	Jan.	Feb.	Mar.	Apr.	May	Jun.	Jul.	Aug.	Sep.	Oct.	Nov.	Dec.	HDD
2014				4.2	9.3	12.6	20	16.5	11.5	5	1	−0.7	3926
2015	−3.1	−0.2	0.1	4	8.4	12.1	15.2	16.5	12.1	5.9	3.9	1.2	3546
2016	−9.7	−2.3	0.1	3.3	11	14	17.2	14.6	11.9	4	−1.1	−0.5	4124

The modeling of the thermophysical properties of pavement materials on the evolution of temperature depth profiles in different climatic regions has been refereed by Hall et al. [19].

2. Materials and Methods

According to Strzelczyk et al. [20] it is important to know the thermal parameters of the surroundings, the weather, and the solar radiation properties in a given location in order to design and improve energy systems. The shallow geothermal energy originates mainly from sun. The measurements in this study were obtained by three main sources: (1) a pyranometer, (2) a heat flux plate, and (3) a distributed temperature sensing (DTS) system. These three independent methods complement each other, giving on a reliable picture of the different layers. The pyranometer measured solar irradiance, the heat flux plate measured the heat flux through the asphalt layer, and the DTS described the temperature distribution down to several meters' depth.

2.1. Pyranometer Measurements

In this research solar irradiance is measured by using a Hukseflux LP02-TR (Hukseflux Inc., Delft, The Netherlands) pyranometer [21] which was placed on the roof of the neighbour building. In this position it was under the open sky without shadows. The measured values were sampled using a DataTaker—data logger (Thermo Fisher Scientific Australia Pty Ltd.,Melbourne, Australia) in 10 s time intervals. The data was logged with a timestamp and stored for analysis. The calibration accuracy of the pyranometer is less than 1.8%.

2.2. Heat Flux Measurements on the Asphalt Surface

The heat flux plate together with pyranometer can be used to study the efficiency of an asphalt layer as a heat collector. Heat flux data were collected using a Hukseflux heat flux plate HFP01-15 (Hukseflux Inc, The Netherlands) [22] buried 5 cm below the asphalt surface at the data collection site (see Figures 2 and 3). This sensor plate generates a small analog output voltage proportional to the net heat flux passing through it. The heat flux towards the ground is interpreted as positive and the heat flux from the ground to the surface is interpreted as negative. The aim of taking the heat flux measurements is to quantify the net energy from solar radiation absorbed by the asphalt during the day and the amount of thermal energy released back to the atmosphere during the night. These energy flows depend on multiple factors, such as irradiance, air temperature, the weather, current soil temperature, the thermal properties of the medium, thermal conductivity, and heat capacity [23]. The accuracy of calibration is ±3%.

A data collection system was developed for reading, transferring, and storing the ground heat flux plate data. This system consisted of an analog-to-digital converter (ADC) and a wireless sensor platform, a wireless gateway, and an embedded PC (see Figure 4). The analog voltage provided by the sensor was digitized by an ADC and the wireless sensor platform sampled the value every 10 s and sent the values to the wireless gateway platform located inside the building. The wireless gateway was connected to the embedded PC, which stored the data on the hard drive. During the data collection period, the measurement devices faced cold temperatures, as low as −30 °C, and continued to operate.

Figure 2. Installation of heat flux plate "Hukseflux" before asphalt pavement.

Figure 3. Heat flux data collection site, "Hukseflux," under the plank.

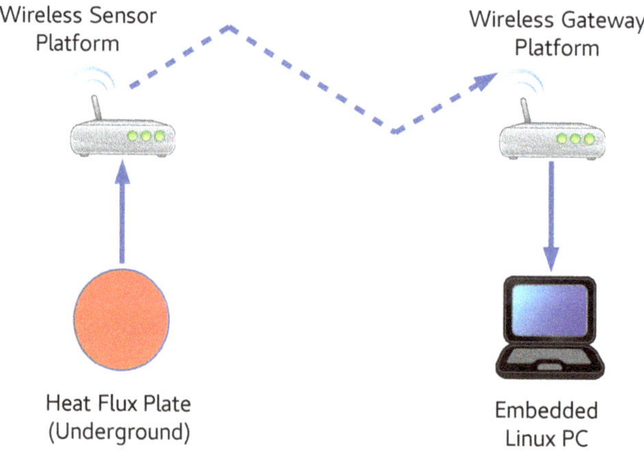

Figure 4. Network structure for wireless data collection of the heat flux sensor.

The collected data were analyzed using the mathematical computing software MATLAB (Version 9.2., The MathWorks Inc., Natick, Massachusetts, USA) to extract information about the net flux of the heat depending on the meteorological conditions of the season and the time of day. Statistical computations were used to make the estimations of the future heat transfers.

2.3. Underground Temperature Measurements

The distributed temperature sensing (DTS) method was used to measure the temperatures in soil layers under the asphalt pavement. This method was chosen because it is known to be usable in cold climate regions [24] and is successfully used in monitoring of boreholes too. The boreholes can be open-holes or sealed wells, as in this study, and the installation can be permanent [25–27]. This method is based on optical light scattering in fiber [28]. Short pulses of laser light are sent to the optical fiber by the DTS measurement device. Part of the incident light pulse is scattered by elastic scattering while it moves along the core of the fiber. The properties of the scattered light are acquired by the DTS measurement device, which then estimates the temperature based on the temperature dependent part of elastic scattering [29,30]. An optical fiber, therefore, can be used as a linear sensor. In this study, the temperatures were observed with 1 m spatial resolution.

A measurement fiber was installed into the vertical boreholes drilled under the asphalt (more exactly described by Mäkiranta et al. [4]). The aim was to measure soil temperatures from the depth of 50 cm down to 10 m at the same time. Temperatures under the asphalt layer were acquired by a DTS device (Oryx DTS, Sensornet Ltd, Hertfordshire, UK) once or twice per month, depending on the season. These measurements were made on site using special instruments in addition to the meteorological data. The accuracy of the DTS device is +/−0.5 °C.

The structure of an asphalt-paved parking lot is shown in Figure 5. The thickness of the asphalt layer is between 7 and 10 cm, the gravel and sand layer is about 60 cm thick , and the clay layer lies beneath those. The physical properties of soil is listed in Table 2.

Figure 5. Soil layers under the asphalt in the measurement field at the University of Vaasa campus area.

Table 2. Some physical properties of the soil in the literature [31–35].

Soil Type	Specific Heat Capacity [kJ/kg·°C]	Density [kg/m^3]
Asphalt	0.92	2400
Gravel	1.50 (dry)	1680 (dry)
Sand	0.84 (dry, 20 °C)	2660
Clay	0.88 (10% moisture) 1.76 (50% moisture)	1600 (dry) 1760 (wet)

3. Results

3.1. Analyzing Pyranometer and Heat Flux Data

The ratio of the incoming solar irradiance leading to a heat flux under the asphalt can be used as a measure of the efficiency of the asphalt layer as a heat collector. The daily average absorption ratio, σ, for the given day was calculated by integrating the solar irradiance, E_e (radiant exposure [H] = Wh/m^2) and heat flux, ϕ, over a 24 h period, as follows:

$$\sigma = \frac{(1/T)\int_{t\in T}\phi(t)\mathrm{dt}}{(1/T)\int_{t\in T}E_e(t)\mathrm{dt}} = \frac{\overline{\phi}}{\overline{E_e}} \quad (1)$$

This formula applied for the 23th of April gives the following result:

$$\sigma = \frac{14\ \mathrm{Wh/m^2}}{230\ \mathrm{Wh/m^2}} = 5.9\%, \quad (2)$$

As expected, the soil warms up in April, which is spring time in the northern hemisphere. The net heat flux is rather small due to high negative flux during the night. Liquid circulation assisted by pumps could be utilized for optimizing the heat transfer to the soil. Heat loss can also be reduced by many other technical means, such as by selecting optimal materials in different soil layers, or by using an insulating cover during the night. The full potential of the asphalt layer can be estimated by calculating the absorption ratio in an optimal case where the nighttime heat loss is totally eliminated by accumulating only the positive heat flux. The absorption ratio of this optimal case, σ_p, is:

$$\sigma_p = \frac{139\ \mathrm{Wh/m^2}}{230\ \mathrm{Wh/m^2}} = 60\%, \quad (3)$$

Although the average heat flux was 139 Wh/m^2 during the daytime, which is 60% of the daytime solar irradiance, the absorption rate for the entire day was only 14 Wh/m^2, which is only 5.9% of the solar irradiance. This reduction was caused by the heat loss that occurred during the nighttime; see Figure 6.

3.2. Relation between the Cumulative Heat Flux and Soil Temperature

The soil temperature, T, is directly proportional to the heat, Q, stored in the soil since $Q = c\,m\,\Delta T$, where m is the mass of the volume unit of soil $[m]$ = kg/m^3 and $[c]$ = [kJ/kg·°C] is the specific heat capacity of soil. Since the heat, Q, is an integral of heat flux, ϕ, in time t and the surface S, the following equation, showing the dependency between soil temperature and heat flux, is obtainable:

$$\int_0^t \oint_S \vec{\phi}\cdot d\vec{s}\,dt = c\,m\,\Delta T, \quad (4)$$

where ϕ is the heat flux into the unit of soil volume surrounded by the closed surface, S, during time dt.

The pyranometer and the heat flux plate data were used for estimating the absorption efficiency of the asphalt layer. Changes in the underground temperatures were acquired using the DTS measurement method, and they are proportional to the integral sum of the net heat flux.

As an example of daily data analysis, Figure 6 represents the measured solar irradiance and heat flux on the 23rd of April.

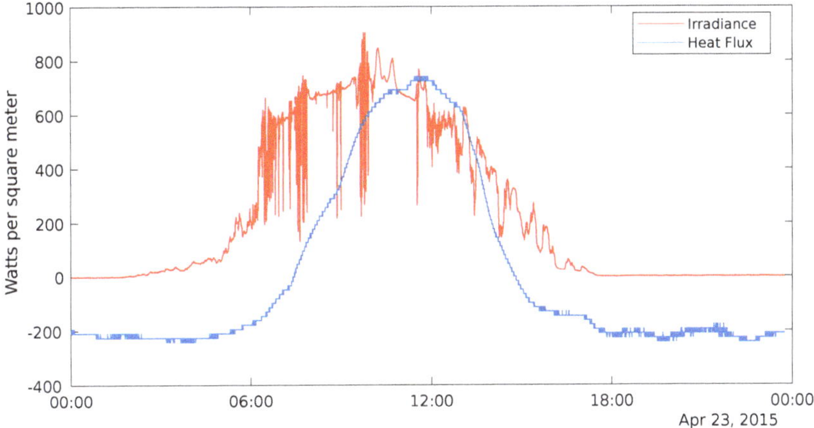

Figure 6. Solar irradiance, $E_e(t)[\mathrm{W/m^2}]$ (pyranometer) and heat flux, $\overline{\phi}$ $[\mathrm{W/m^2}]$ (heat flux plate), as a function of time (h) for the 23rd of April 2015.

The solar irradiance measured by the pyranometer is always positive since it represents the solar energy received as watts per square meter. The data shown in Figure 6 spans from 00:00 until 23:59. The sunlight on the 23rd of April 2015 started at 3:30 due to the northern latitude of Finland and lasted until 17:00, having a peak of 900 $\mathrm{W/m^2}$ at 11:00. The data were sampled every 10 s and included relatively large and fast fluctuations due to cloud shadows and reflections.

Figure 6 also represents the net heat flux, ϕ, through the asphalt surface for the same day. Heat flux can have negative values when the thermal energy is dissipated back onto the surface from the ground. The figure shows that the solar irradiance induces a positive heat flux towards the ground, whereas the heat flux becomes negative when the irradiance is close to zero. The solar irradiance must be greater than the heat loss to achieve positive heat flux. The heat loss depends on the soil temperature, air temperature, and the weather conditions. The heat loss is relatively small in spring when the soil is still cool and relatively high in autumn when the soil is warm. Although the peak heat flux was 700 $\mathrm{W/m^2}$, the average heat flux ϕ for the given day was only 14 $\mathrm{W/m^2}$. The average heat loss was approximately 130 $\mathrm{Wh/m^2}$.

Figures 7–10 illustrate the instantaneous balances during 5 day periods of heat flux (average net heat flow and average positive heat flow are measured with heat flux plate, including diffuse irradiance) and irradiance (direct solar irradiance measured by pyranometer) in four different seasons. The data shown in the figures are summarized in Table 3. Heat flux value for autumn is negative because the soil is still warm but is cooling down continuously. In winter, the positive heat flux is small, depending more on the air temperature than negligible irradiance, and losses are small due to the annual soil temperature minimum and the insulating snow cover. The soil starts warming quickly during spring when the irradiance is increasing but the soil temperature is still low, keeping losses small. Solar irradiance is greatest during the summer, but the net heat flux is not very strong because the soil is already warm and the losses are also high.

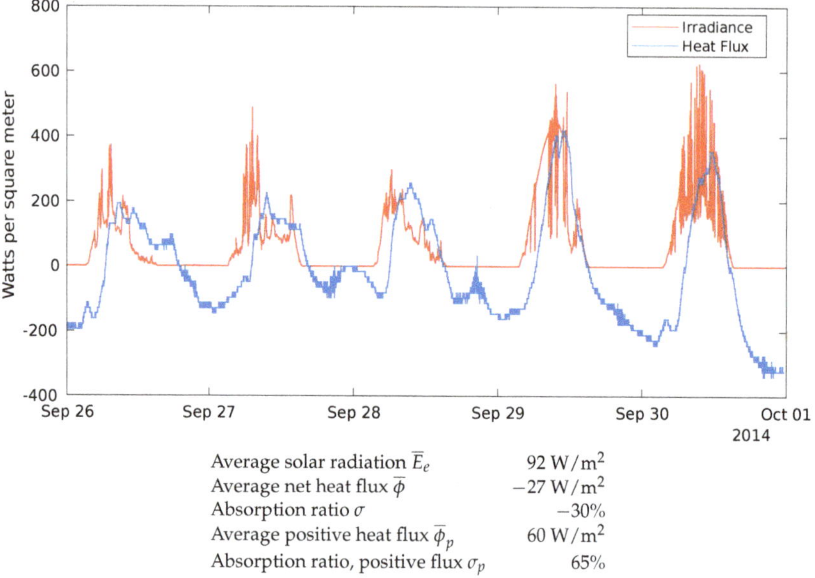

Figure 7. Instantaneous solar irradiation $\overline{E}_e$ [W/m^2] and heat flux $\overline{\phi}$ [W/m^2] from 26 Sep to 01 Oct 2014 (autumn) as a function of time and estimated average values as a function of time (day).

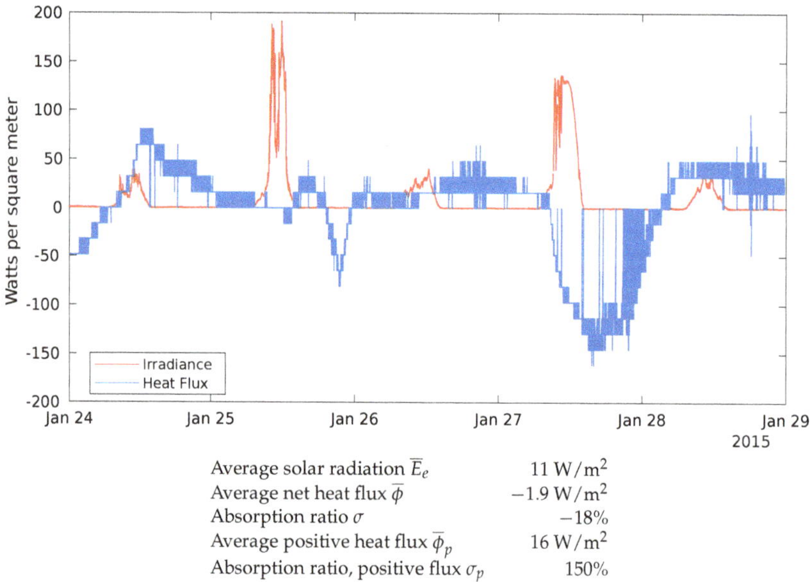

Figure 8. Instantaneous solar irradiation [W/m^2] and heat flux [W/m^2] from 24 to 29 Jan 2015 (winter), and their estimated average values as a function of time (day). Both irradiance and heat flux are negligible and heat flux is more dependent on the temperature than irradiance.

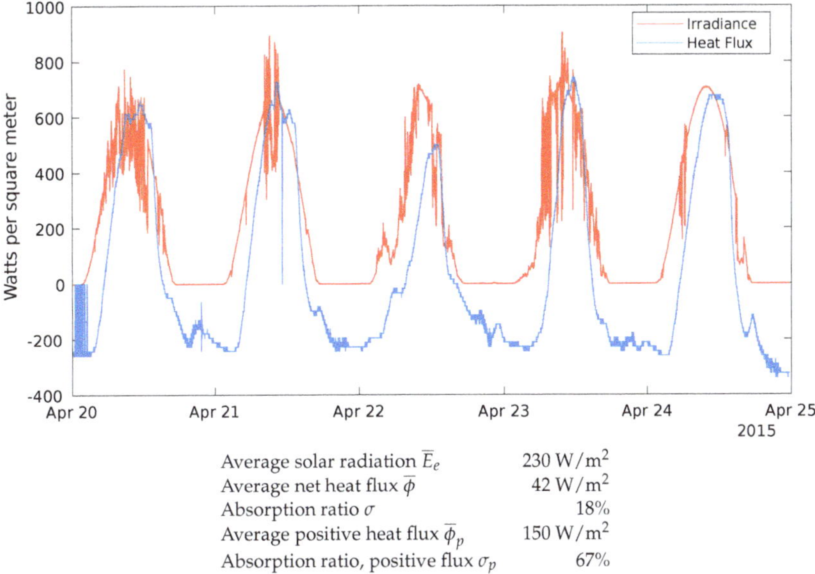

Average solar radiation $\overline{E}_e$	230 W/m^2
Average net heat flux $\overline{\phi}$	42 W/m^2
Absorption ratio σ	18%
Average positive heat flux $\overline{\phi}_p$	150 W/m^2
Absorption ratio, positive flux σ_p	67%

Figure 9. Instantaneous solar irradiation $\overline{E}_e$ [W/m^2] and heat flux [W/m^2] from 20 to 25 Apr 2015 (spring) as function of time (day) and their estimated average values.

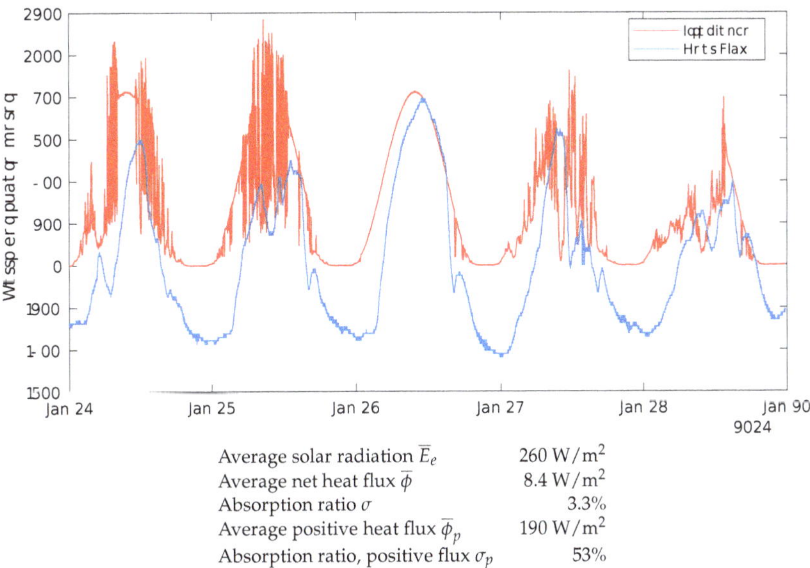

Average solar radiation $\overline{E}_e$	260 W/m^2
Average net heat flux $\overline{\phi}$	8.4 W/m^2
Absorption ratio σ	3.3%
Average positive heat flux $\overline{\phi}_p$	190 W/m^2
Absorption ratio, positive flux σ_p	53%

Figure 10. Instantaneous solar irradiation $\overline{E}_e$ [W/m^2] and heat flux $\overline{\phi}$ [W/m^2] from 15 to 20 Jun 2015 (summer) as a function of time (day) and their estimated average values.

Because thermal energy loss due to radiation and convection also occurs during daytime, the positive heat flux could be further improved by lowering the temperature of the surface during daytime hours; for example, by collecting and transferring thermal energy to seasonal storage. This would allow the utilization of this urban renewable energy even in cold climate regions. Further research is needed to find the full potential of the asphalt heat collection and storage system.

3.3. Temperature Distribution Measurements Using DTS

Using the DTS system, temperatures were measured periodically at certain depths. These measurements helped to analyze how the temperature below the surface is distributed. The selected measurement depths in this research are 0.5 m, 1 m, 1.5 m, 3 m, 5 m, and 10 m. Measurements for a period of one year are represented in Figure 11. From April to September, temperatures rise in the shallowest layers. Instead, from October to March the temperatures decrease. At the depth of 10 m the differences in temperature in the shallowest layers have no effect anymore.

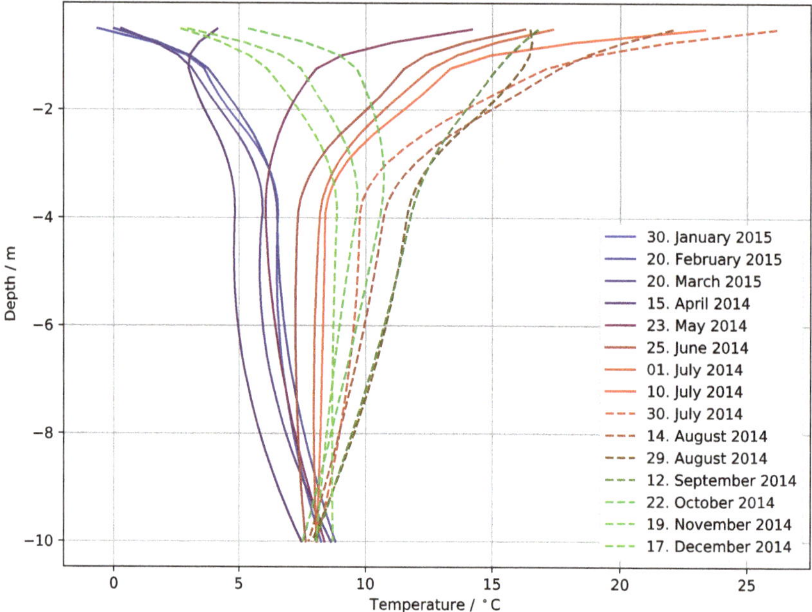

Figure 11. Seasonal soil temperatures at different depths from the surface according to the DTS measurements. During the summer (April–September) temperatures near the surface increase, and during winter months (October–March) the surface is the coldest. At 10 m depth the effect of the Sun is diminished.

Figure 12 represents the temperature changes of the upper layers of the ground from the surface down to 10 m. The data, acquired from the DTS system, clearly indicates that the deeper layers of the soil react to the heat flux with delay compared to the shallowest layers of the soil. This delay depends on soil type and heat conductivity, k. The curves for July and October show that the shallowest layers are already cooled down in October, but the temperatures of depths below 4 m tend to stay steady. The curves for October and December show that while the shallowest layers keep cooling down, the temperatures of deeper layers also drop. After the snow melts in April, the temperature of the shallowest layer begins to increase but the deeper layers are not affected. During May and the following months of summer, the temperatures continue to increase. Deeper layers are always more steady than the surface, and they absorb or release thermal energy much more slowly. The temperature change near the surface is positive between April and August and negative from the beginning of August until the beginning of March. The deeper layers of soil keep the thermal energy longer than the surface layer, and therefore, the cumulative thermal energy begins decreasing in the middle of October. When fast temperature increase occurs on the shallowest layers, some thermal energy can be delivered to the deeper layers to be stored.

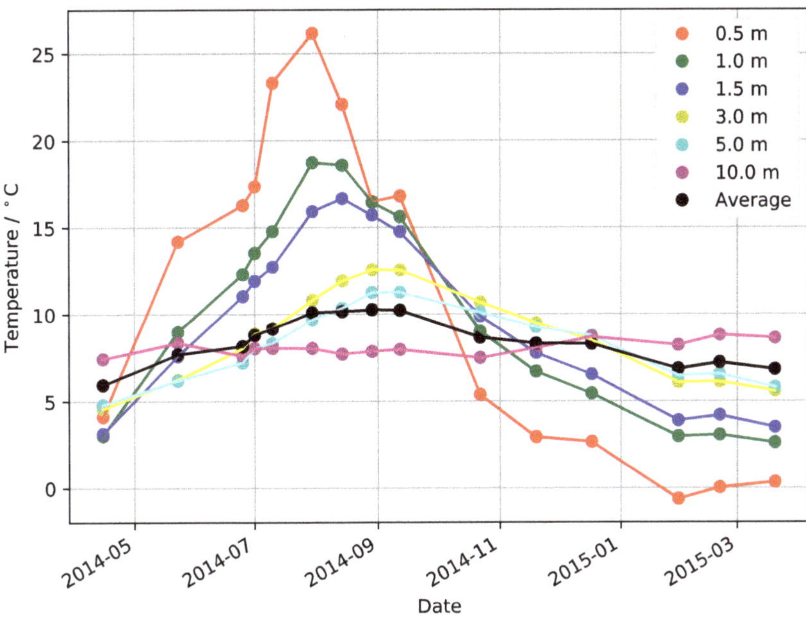

Figure 12. Soil temperatures measured at different depths (0.5 m, 1.0 m, ..., 10 m) and the weighted average temperature of the whole 10 m deep layer.

Figure 13 shows the solar irradiation and the cumulative heat flux from asphalt to the deeper soil layers. In addition to showing the net heat flux, the negative and positive parts of the heat flux are also integrated separately to obtain a better understanding of the flow. The cumulative net flow and the temperature of the soil are increasing until the 26th of September, when the cumulative irradiance is 870 kWh/m^2; cumulative positive, negative, and net flows are 610 kWh/m^2, 460 kWh/m^2, and 150 kWh/m^2, respectively. The positive, negative, and net flows correspond to 70%, 53%, and 17%, respectively, of the cumulative solar irradiance.

Figures 12 and 13 are not directly comparable, since they contain measurements taken in different years, but show similar results. Furthermore, the heat flux of the asphalt surface is not directly related to the internal heat exchanges inside the soil. Better conformance is obtained by comparing the cumulative heat flux with the average temperature through all soil layers, shown by the black line in Figure 12.

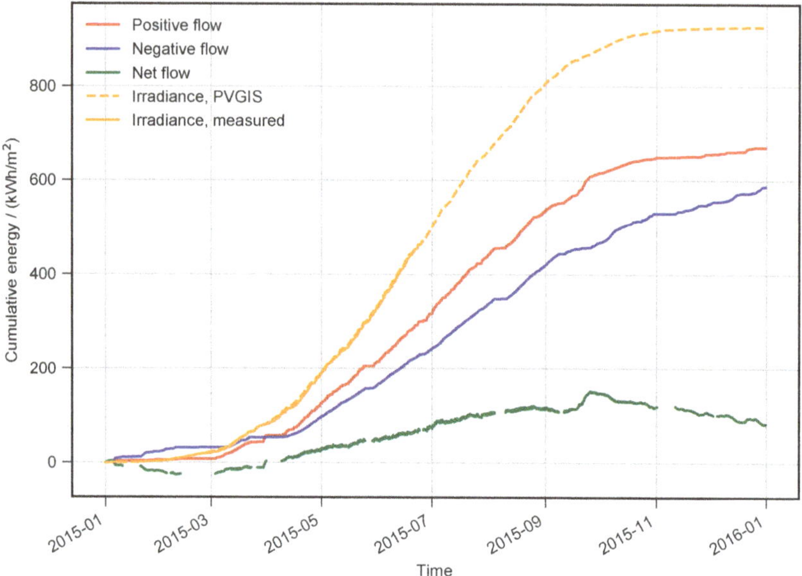

Figure 13. Hourly integrated cumulative irradiance and cumulative heat flux under the asphalt layer over an annual cycle. The net flow consists of the cumulative hourly positive flow into the ground and the cumulative hourly negative flow escaping from the ground. After the warming period, in 26th of September the positive flow is about 70% of the solar irradiance and the net flow is only about 17% of it. Irradiance PVGIS is the Solar irradiance estimated using the Photo Voltaic Geographical Information System [36,37].

4. Discussion and Conclusions

The data collected in this research make it possible to analyze the performance of the asphalt as a heat collection system during any time of the year. Selected periods of five-day data from every season of the year were analyzed and are represented in Figures 7–10.

When thermal energy is collected by the asphalt layer, the temperature rises, causing a temperature gradient that drives a heat flux through the soil, according to Fourier's law, $\phi = -k\nabla T$. Here k is the thermal conductivity $[k]$ = W/m K, and T is the temperature $[T]$ = K . The temperature distributions shown in Figure 11 reveal a high gradient in the topmost gravel layer and a gentler sloping gradient in the deeper layers, such as clay and bedrock. The same observation was seen in the preceding laboratory measurements as well, as reported in [14]. This structure is not desirable for solar energy storage. The thermal energy flowing through the asphalt is not efficiently distributed in the bigger soil volume because it is dissipated back to the atmosphere during the night until the thermal energy penetrates deeper layers (Figure 13). The heat loss can be eliminated by decreasing the soil surface temperature close to the ambient temperature by transferring the thermal energy from the surface. For this purpose, the model in which the energy is harvested from the surface and stored 6 m under the surface, expressed by Ho et al. [15], could be a working solution. The ground structure could be optimized by increasing the conductivity of the surface by changing the materials or by irrigation. According to Loomans et al. [6] the thermal energy could be harvested inside the asphalt, between two asphalt layers. In their model there are several layers—a thin asphalt layer on the surface, the asphalt pavement layer, and an asphalt layer with heat transfer tubes, and beneath them, asphalt layers again.

Figures 12 and 13 indicate that the net heat flux becomes positive during March or April, depending on the year, and turns back to negative in September or October. The surface reaches its peak temperature at the end of July but remains warm enough to continue driving positive net heat flux to the ground until the 26th of September. The average net heat flux is less than 15% of the available irradiance due to the nighttime thermal energy losses, while the average positive thermal energy is 64% of the irradiance. This positive heat flux could be more efficiently utilized by reducing nighttime losses, which were, according to Table 3, 3/4 of the positive heat flux during spring and summer, and even higher during autumn.

Table 3. Average solar irradiance, heat flux, and absorption ratios over five day periods in all seasons. The first three rows represent the average net heat collection values over the whole period and the next two rows represent the corresponding values for positive values.

Parameter	Autumn	Winter	Spring	Summer	Yearly Average
Average solar irradiance $\overline{E}_e$	92 W/m^2	11 W/m^2	230 W/m^2	260 W/m^2	148 W/m^2
Average net heat flux $\overline{\phi}$	−28 W/m^2	−1.9 W/m^2	42 W/m^2	8.4 W/m^2	9.5 W/m^2
Absorption ratio σ	−30%	−18%	18%	3.3%	6.4%
Average positive heat flux $\overline{\phi}_p$	60 W/m^2	16 W/m^2	150 W/m^2	190 W/m^2	104 W/m^2
Absorption ratio, positive flux σ_p	65%	145%	65%	73%	70%

The measurements revealed that the current efficiency of the asphalt pavement in absorbing solar irradiation during the soil warming period between 1st of January and 26th of September is about 17% (see Figure 13). If the nighttime escape of thermal energy can be eliminated, transferring thermal energy away from the surface, the efficiency could be up to 70% in the same time period (Table 3). The efficiency could also be increased by covering the asphalt field with insulating cover during the night. The cumulative irradiance, net heat flux, and positive heat flux in the asphalt field during the soil warming period were 870 kWh/m^2, 150 kWh/m^2, and 610 kWh/m^2, respectively. Cumulative irradiance, net heat flux, and positive heat flux during the whole year were 930 kWh/m^2, 83 kWh/m^2, and 670 kWh/m^2, respectively. These results imply that the asphalt layer could potentially collect up to 670 kWh/m^2 (see Figure 13), provided that the losses can be properly handled. The temperature of the surface layer is relatively low, exceeding +20 °C at 0.5 m depth only in the middle of the summer. The temperature of the heat collection liquid usually needs to be elevated using heat pumps before it can be used for heating purposes. In some applications, such as melting the snow on the asphalt field or making electricity by using thermocouples [38], it could be used without heat pumps. DTS measurements show (Figure 11) that the annual changes in the soil temperature, caused by the solar radiation, mostly occur within the first 10 m below the ground level. Identical results have been found by Ho et al. [15].

Some unexpected environmental factors occurred while taking the measurements, such as shadows of cars parking over the measurement area. The long time, which were needed to take the measurements, caused some interruption in data acquisition. These interruptions were mainly caused by battery depletion, software resets due to power shortages, and other unknown reasons. Measurement equipment could be developed by getting an instrument amplifier and backup power. In the future, these losses could be estimated and compensated for using Kalman filters and other estimation methods. In such a study, the asphalt and the layers below it are typically used in Finland. These results must not be generalized if the layers and the surroundings differ from those in this research. The data were acquired during a 1–2 year period, which is relatively short term taking into account the annual variation in weather.

In this research, the heat collecting properties of the existing asphalt fields were analyzed and some suggestions for improving the collection efficiency were made. Three independent methods were used to measure the solar irradiance, heat flux, and soil temperature, to explain the behavior of the asphalt pavement for collecting solar energy. This combination of methods was found to give valuable information for harvesting and storing of solar energy in cold climate region. Further research using thermal simulations and experiments is needed to study how much the active and passive heat transfer mechanisms, such as under-asphalt soil structures, could improve the net thermal energy collection efficiency in a practical case. Complementary information could be measured by using a radio net meter to measure the absorption and emission of the asphalt. [6] When the vertical temperature distribution (thermal gradient) in the soil is known, it can be used to optimize geothermal energy harvesting and storing applications in the future. An interesting model with which to apply the results of this study for a snow-melting system was proposed by Ho et al. [15]. The model consists of two pipe lines. The first one is embedded below the pavement and an another is located 6 m under the surface. The upper pipeline can be used to harvest heat in summer and for snow-melting in winter. Another pipeline can be used as a seasonal energy storage in summer and used for snow-melting in winter. The optimized asphalt pavements could act as solar collectors [1,39] and mitigate the urban heat island (UHI) effect [40].

Author Contributions: Software development, data curation, formal analysis C.Ç.; validation, C.Ç. and P.V.; methodology, investigation, writing–original draft preparation, visualization, C.Ç. and A.M.; resources, E.H.; writing–review and editing, A.M. and E.H.; supervision, P.V., E.H. and M.E. All authors have read and agreed to the published version of the manuscript.

Funding: This research received no external funding.

Acknowledgments: The authors gratefully acknowledge the financial support of the City of Vaasa and the Graduate School of the University of Vaasa.

Conflicts of Interest: The authors declare no conflict of interest.

References

1. Bobes-Jesus, V.; Pascual-Muñoz, P.; Castro-Fresno, D.; Rodriguez-Hernandez, J. Asphalt solar collectors: A literature review. *Appl. Energy* **2013**, *102*, 962–970. [CrossRef]
2. Bijsterveld, W.T.V.; Houben, L.J.M.; Scarpas, A.; Molenaar, A.A.A. Using Pavement as Solar Collector: Effect on Pavement Temperature and Structural Response:. *Transp. Res. Rec.* **2001**. [CrossRef]
3. Li, C.; Shang, J.; Cao, Y. Discussion on energy-saving taking urban heat island effect into account. In Proceedings of the 2010 International Conference on Power System Technology, Hangzhou, China, 24–28 October 2010; pp. 1–3. [CrossRef]
4. Mäkiranta, A.; Hiltunen, E. Utilizing Asphalt Heat Energy in Finnish Climate Conditions. *Energies* **2019**, *12*, 2101. [CrossRef]
5. Majorowicz, J.; Grasby, S.E.; Skinner, W.R. Estimation of Shallow Geothermal Energy Resource in Canada: Heat Gain and Heat Sink. *Nat. Resour. Res.* **2009**, *18*, 95–108. [CrossRef]
6. Loomans, M.; Oversloot, H.; De Bondt, A.; Jansen, R.; Van Rij, H. Design tool for the thermal energy potential of asphalt pavements. In Proceedings of the Eighth International IBPSA Conference, Eindhoven, The Netherlands, 11–14 August 2003; pp. 745–752.
7. Hailu, G.; Hayes, P.; Masteller, M. Long-Term Monitoring of Sensible Thermal Storage in an Extremely Cold Region. *Energies* **2019**, *12*, 1821. [CrossRef]
8. Mehrpooya, M.; Hemmatabady, H.; Ahmadi, M.H. Optimization of performance of Combined Solar Collector-Geothermal Heat Pump Systems to supply thermal load needed for heating greenhouses. *Energy Convers. Manag.* **2015**, *97*, 382–392. [CrossRef]
9. Hesaraki, A.; Holmberg, S.; Haghighat, F. Seasonal thermal energy storage with heat pumps and low temperatures in building projects—A comparative review. *Renew. Sustain. Energy Rev.* **2015**, *43*, 1199–1213. [CrossRef]
10. Sibbitt, B.; McClenahan, D.; Djebbar, R.; Thornton, J.; Wong, B.; Carriere, J.; Kokko, J. The Performance of a High Solar Fraction Seasonal Storage District Heating System—Five Years of Operation. *Energy Procedia* **2012**, *30*, 856–865. [CrossRef]
11. Lund, J.W.; Freeston, D.H.; Boyd, T.L. Direct utilization of geothermal energy 2010 worldwide review. *Geothermics* **2011**, *40*, 159–180. [CrossRef]
12. Mesquita, L.; McClenahan, D.; Thornton, J.; Carriere, J.; Wong, B. Drake Landing Solar Community: 10 Years of Operation. In *Proceedings of SWC2017/SHC2017*; International Solar Energy Society: Abu Dhabi, UAE, 2017; pp. 1–12. [CrossRef]

13. Haq, H.M.K.U.; Hiltunen, E. An inquiry of ground heat storage: Analysis of experimental measurements and optimization of system's performance. *Appl. Therm. Eng.* **2019**, *148*, 10–21. [CrossRef]
14. Martinkauppi, J.B.; Mäkiranta, A.; Kiijärvi, J.; Hiltunen, E. Thermal Behavior of an Asphalt Pavement in the Laboratory and in the Parking Lot. *Sci. World J.* **2015**, *2015*. [CrossRef]
15. Ho, I.H.; Dickson, M. Numerical modeling of heat production using geothermal energy for a snow-melting system. *Geomech. Energy Environ.* **2017**, *10*, 42–51. [CrossRef]
16. File: Fennoscandia.png—Wikimedia Commons, the Free Media Repository. 2019. Available Online: https://commons.wikimedia.org/w/index.php?title=File:Fennoscandia.png&oldid=365664646 (accessed on 14 January 2020).
17. Finnish Meteorological Institute FMI: Heating Degree Days. Available online: https://en.ilmatieteenlaitos.fi/heating-degree-days (accessed on 13 January 2020).
18. Finnish Meteorological Institute FMI: Monthly Average Air Temperature Data 2014–2016, Klemettilä, Vaasa. Available online: https://ilmatieteenlaitos.fi/havaintojen-lataus#! (accessed on 28 January 2019).
19. Hall, M.R.; Dehdezi, P.K.; Dawson, A.R.; Grenfell, J.; Isola, R. Influence of the Thermophysical Properties of Pavement Materials on the Evolution of Temperature Depth Profiles in Different Climatic Regions. *J. Mater. Civ. Eng.* **2012**, *24*, 32–47. [CrossRef]
20. Strzelczyk, P.; Szewczyk, M.; Gałek, R.; Gil, P. Measurement of solar radiation properties and thermal energy of the atmosphere in Rzeszow. In *Zeszyty Naukowe Politechniki Rzeszowskiej. Mechanika*; z. 90 [298], nr 4; Rzeszow University of Technology: Rzeszow, Poland, 2018.
21. Van den Box, K. Hukseflux User Manual LP02, Second Class Pyranometer. 2015. Available online: https://www.hukseflux.com/uploads/product-documents/LP02_manual_v1606.pdf (accessed on 12 September 2019).
22. Hoeksema, E. User Manual, HFP01 & HFP03, Heat Flux Plate/Heat Flux Sensor. 2015. Available Online: https://www.hukseflux.com/uploads/product-documents/HFP01_HFP03_manual_v1721.pdf (accessed on 12 September 2019).
23. Zhu, K.; Blum, P.; Ferguson, G.; Balke, K.D.; Bayer, P. The geothermal potential of urban heat islands. *Environ. Res. Lett.* **2010**, *5*, 044002. [CrossRef]
24. Fisher, A.T.; Mankoff, K.D.; Tulaczyk, S.M.; Tyler, S.W.; Foley, N.; WISSARD Science Team. High geothermal heat flux measured below the West Antarctic Ice Sheet. *Sci. Adv.* **2015**, *1*, e1500093. [CrossRef]
25. Ouyang, L.B.; Belanger, D. Flow Profiling via Distributed Temperature Sensor (DTS) System—Expectation and Reality. In Proceedings of the Society of Petroleum Engineers, SPE Annual Technical Conference and Exhibition, Houston, TX, USA, 26–29 September 2004. [CrossRef]
26. Coleman, T.I.; Parker, B.L.; Maldaner, C.H.; Mondanos, M.J. Groundwater flow characterization in a fractured bedrock aquifer using active DTS tests in sealed boreholes. *J. Hydrol.* **2015**, *528*, 449–462. [CrossRef]
27. Henninges, J.; Zimmermann, G.; Büttner, G.; Schrötter, J.; Erbas, K.; Huenges, E. Fibre-optic temperature measurements in boreholes. In Proceedings of the 7th FKPE-Workshop "Bohrlochgeophysik and Gesteinsphysik", GeoZentrum Hannover, Hannover, Germany, 23–24 October 2003; pp. 23–24.
28. *Oryx DTS User Manual v4*; Sensornet Ltd.:, London, UK, 2007.
29. Ukil, A.; Braendle, H.; Krippner, P. Distributed Temperature Sensing: Review of Technology and Applications. *IEEE Sens. J.* **2012**, *12*, 885–892. [CrossRef]
30. Sayde, C.; Gregory, C.; Gil-Rodriguez, M.; Tufillaro, N.; Tyler, S.; Giesen, N.V.d.; English, M.; Cuenca, R.; Selker, J.S. Feasibility of soil moisture monitoring with heated fiber optics. *Water Resour. Res.* **2010**, *46*. [CrossRef]
31. Xu, Q.; Solaimanian, M. Modeling temperature distribution and thermal property of asphalt concrete for laboratory testing applications. *Constr. Build. Mater.* **2010**, *24*, 487–497. [CrossRef]
32. Valtanen, E. *The Handbook of Math and Physics, Matematiikan ja Fysiikan Käsikirja*, 2 ed.; Genesis-kirjat Oy: Jyväskylä, Finland, 2007.
33. Engineering Toolbox. Densities and Heat Capacities of Materials. Available Online: https://www.engineeringtoolbox.com (accessed on 15 December 2019).
34. Blocon, A. EED Earth Energy Designer-Vertical Borehole Design Program for. 2015. Available online: https://www.buildingphysics.com/manuals/EED3.pdf (accessed on 20 October 2019).
35. Ronkainen, N. *Properties of Finnish Soil Types*; Finnish Environment Institute: Helsinki, Finland, 2012. Available online: https://core.ac.uk/download/pdf/14927376.pdf (Accessed on 5 December 2019).

36. Gracia, A.M.; Huld, T.; European Commission; Joint Research Centre; Institute for Energy and Transport. *Performance Comparison of Different Models for the Estimation of Global Irradiance on Inclined Surfaces: Validation of the Model Implemented in PVGIS*; Publications Office: Luxembourg, 2013.
37. European Comission, J.R.G. Photovoltaic Geographical Information System. 2014. Available Online: http://re.jrc.ec.europa.eu/pvgis (accessed on 21 September 2019).
38. Datta, U.; Dessouky, S.; Papagiannakis, A.T. Thermal Energy Harvesting from Asphalt Roadway Pavement. In *Advancement in the Design and Performance of Sustainable Asphalt Pavements*; Mohammad, L., Ed.; Sustainable Civil Infrastructures; Springer International Publishing: Cham, Switzerland, 2018; pp. 272–286. Available online: https://link.springer.com/chapter/10.1007/978-3-319-61908-8_20 (accessed on 20 February 2020)
39. Basheer Sheeba, J.; Krishnan Rohini, A. Structural and Thermal Analysis of Asphalt Solar Collector Using Finite Element Method. *J. Energy* **2014**, *2014*, 602087. [CrossRef]
40. Pascual-Muñoz, P.; Castro-Fresno, D.; Serrano-Bravo, P.; Alonso-Estébanez, A. Thermal and hydraulic analysis of multilayered asphalt pavements as active solar collectors. *Appl. Energy* **2013**, *111*, 324–332. [CrossRef]

MDPI
St. Alban-Anlage 66
4052 Basel
Switzerland
Tel. +41 61 683 77 34
Fax +41 61 302 89 18
www.mdpi.com

Energies Editorial Office
E-mail: energies@mdpi.com
www.mdpi.com/journal/energies

www.ingramcontent.com/pod-product-compliance
Lightning Source LLC
LaVergne TN
LVHW070652170726
843515LV00004B/390

* 9 7 8 3 0 3 9 3 6 2 8 4 4 *